Creating a Sea Change

The WWF/IUCN Marine Policy

WWF/Thierry Petit/BIOS

Acknowledgements

The WWF/IUCN Marine Policy has evolved over several years and thanks are due to many people. Don Hinrichsen prepared the first draft of the text, which has been revised by Sue Wells. The policy itself is the result of input from past and present members of the WWF Marine Advisory Group and associated experts. Particular thanks go to the following people for their help:

J. Beckett (Consultant, Canada)
C. Bijasson (WWF International)
J. Clay (WWF-US)
C. von Dorrien (WWF-Germany)
P. Dugan (IUCN-HQ)
B. Eichbaum (WWF-US)
C. Flockhart (WWF International)
C. de Fontaubert (IUCN-US)
M. Gawler (WWF International)
B. Groombridge (World Conservation Monitoring Centre)
P. Holthus (Marine Aquarium Council)
M. Jorge (LA/C Programme – WWF-US)
D. Karavellas (WWF-Greece)
L.C. Lim (WWF-Malaysia)
C. Lee (WWF-Hong Kong)
I. Lutchman (Consultant, WWF-UK)
S. Lutter (WWF-N.E. Atlantic Programme)
J. McNeely (IUCN-HQ)
E. Meltzer (Consultant, Canada)
M. Mercer (IUCN-Canada)
K. de Meyer (Bonaire Marine Park)
M. Ngoile (University of Dar es Salaam)
R. Nias (WWF-Australia)
C. Phillips (WWF Antarctic and Cetacean Officer)
S. Pullen (WWF-UK)
A. Raselimanana (WWF-Madagascar Programme Office)
C. Recchia (Consultant, WWF International)
A. Rosser (IUCN/SSC)
R. Salm (IUCN-East Africa)
E. Salter (WWF International)
A. Schumm (WWF-Germany)
G. Shepherd (WWF International)
M. Sutton (WWF Endangered Seas Campaign)
D. Tarte (IUCN Councillor)
P. Tous (IUCN-Guinea Bissau Office)
F. Vorhies (IUCN-HQ)
J. Waugh (IUCN-US)

Contents

WWF/Thomas Schuppisser

The untamed quality of the oceans belies their vulnerability to overexploitation by humankind.

Introduction

HUMANS are terrestrial beings and our relationship with the oceans has always been uneasy, based largely on ignorance. The oceans' vastness – they cover over 70 per cent of the planet – and their hidden depths evoke awe and fear. Oceans can be dangerous places for humans, epitomized by the sinking of the *Titanic* – at its time, the world's largest ship and considered unsinkable. Fear of the sea has given rise to the many traditional beliefs in sea monsters, such as the *kraken*, which would entwine whole ships in its tentacles, and has created modern monsters out of sharks through movies such as *Jaws*.

Conquering the sea was, and in many ways still is, an act of heroism. Columbus's voyage across the Atlantic is still celebrated. That of the Chinese admiral Zheng He, who travelled to East Africa, also in the 15th century, is less fêted but was an equal achievement. Their equivalents today – those who explore the black watery deeps in submersibles – also take enormous risks. People whose lives are tightly bound up with the sea, such as fishers and mariners, show immense respect for it. In Hawaii, the shark is known as an *aumakua* or 'guardian god of a family', and many fishing families around the world pray to their gods for protection before any of them sets out on the water.

But ashore, ignorance of the oceans has generated an attitude towards their use which is careless and lacking in respect. We treat them as an infinite food supply, a bottomless pit for waste, and a common space on which to play and fight, available to all. Industrial and domestic development on coasts, overexploitation of marine resources, and burgeoning tourism all compete and interfere with the species and spaces of coasts, the open seas, and the deep oceans.

Although we may have conquered much of the ocean, our long-cherished concept of the 'freedom of the seas' has been challenged. Indeed, we will probably have to recognize that free access to marine resources is a thing of the past. At the end of the 20th century, we can no longer use ignorance as an excuse for mismanagement. The evidence is compelling, in the crowded beaches of the Mediterranean, the dwindling fish catches of the Northwest Atlantic, and the plastic debris washed up on remote and uninhabited Pacific islands.

WWF–Worldwide Fund For Nature and IUCN–The World Conservation Union have therefore joined forces in this publication to outline a policy that will address the complexity of oceans and coasts, as well as our dependency on them. Nowhere is the challenge of protecting and sustainably managing natural resources more daunting than in the marine environment, the last great source of wild-caught food on this planet. WWF and IUCN call upon partner organizations, governments, multilateral institutions, industries, and individuals to join them in translating this policy into solutions that will result in a sea change in the way we think about and use the oceans.

"The last fallen mahogany would lie perceptibly on the landscape, and the last black rhino would be obvious in its loneliness, but a marine species may disappear beneath the waves unobserved and the sea would seem to roll on the same as always."

G. Carleton Ray, 1988
Ecological diversity in coastal zones and oceans
In: E.O. Wilson and F.M. Peter (eds)
Biodiversity
National Academy Press, Washington D.C.

WWF/Henry Ausloos

"This world is a water world, a planet dominated by its covering mantle of ocean, in which the continents are but transient intrusions of land above the surface of the all-encircling sea."

Rachel Carson, 1951
The Sea Around Us

WWF-Canon/Cassandra Phillips

The polar seas, far from being barren wastes, support enormous amounts of plant and animal life.

1. Our blue planet

From the objectivity of space the Earth is overwhelmingly composed of water. Land, in fact, appears almost as a geological afterthought. The surface of the blue planet is over 70 per cent water, all but 3 per cent of it salt water. Life on Earth first evolved in the primordial soup of ancient seas some 3.5 billion years ago. Today, the seas harbour a rich diversity of habitats and species, many of which await discovery. Marine scientists believe that oceans and coastal areas may contain at least as many species of plants and animals as tropical rainforests.

Oceanus – the world ocean – is vast, covering 360 million square kilometres (km^2) and holding close to 1.4 billion cubic kilometres (km^3) of water[1]. This gigantic body of salt water, which wraps around the planet like an insulating blanket, literally makes life on Earth possible. The oceans are the engines that drive the world's climate, defining weather and storing huge quantities of solar energy in the process. As climatologists have discovered, the oceans absorb and store carbon dioxide from the atmosphere. Since this invisible gas is one of the main climate-changing agents, this makes the oceans, like forests, an important carbon 'sink' that helps to modify human impacts on the global climate.

The oceans are also the liquid heart of the Earth's hydrological cycle – nature's great solar-driven water pump – which causes roughly 430,000 km^3 of water to evaporate from the oceans every year. Of this amount, around 110,000 km^3 fall as freshwater precipitation over land (the rest falling over the sea), replenishing surface and ground waters and eventually completing the cycle by returning to the sea.

Ocean currents – the blue planet's super highways – transfer great quantities of water and nutrients from one place to another. The Gulf Stream, for instance, pushes more water than is carried by all the rivers on Earth from the Gulf of Mexico and the Caribbean across the Atlantic into northern Europe. Where currents converge – for instance, off the west coast of South America, West Africa, and the North Atlantic – upwellings of nutrient-rich bottom water fuel an explosion of marine life, including great quantities of phytoplankton and zooplankton (microscopic plants and animals) that form the basis of the marine food chain.

Marine biodiversity

At the basic taxonomic level – the level of fundamental body plans or phyla – marine animals display much greater diversity and are much more specialized than their land-based relatives. Out of a total of 33 animal phyla, 32 are found in the sea[2]. In fact, 15 of these, including comb-jellies, lamp shells, and starfish, are exclusively marine. As a consequence, the sea is rife with life, harbouring everything from jawless fish to giant squid. Although only about 275,000 marine species have been identified and described, compared to 1.5 million on land, most marine biologists consider this to be the tip of the iceberg[3]: there may be more than 1 million species on coral reefs alone, and perhaps as many as 10 million in the deep ocean basins[4]. Fishes are the most abundant group of vertebrates on Earth in terms of numbers of both species and individuals[5].

The oceans contain the world's largest and smallest animals and virtually everything in between. The largest animal on the planet, the blue whale, can measure up to 35 metres long and weigh up to 200 tonnes[6]. At the other extreme, scientists have identified 22 major types of meiofauna, tiny microscopic animals attached to grains of sand. A handful of wet sand may contain up to 10,000 such animals.

WWF/Nancy Sefton

Ocean creatures exhibit a wide range of survival strategies not found on land. The numerous planktonic life forms drift passively in their watery realm, relying on currents to transport them to new nutritional sources and habitats. Filter feeders – ranging in size from tiny zooplankton to baleen whales and whale sharks – sieve plankton and other floating organisms for food. Others, such as toothed sharks, billfish (marlins, swordfish, sailfish), and killer whales, feed at the top of the food chain.

Still other species that spend long periods of time out of the water, including some crabs, seabirds, and polar bears, depend on the seas for sustenance or as breeding areas. Crabs, for instance, use shallow coastal waters in estuaries, mangrove swamps, and salt marshes to feed and breed. Polar bears and most seabirds are almost entirely dependent on the sea for food.

Many marine species migrate over great distances. Billfish and tuna, for instance, cover thousands of kilometres from feeding to spawning areas and back. Turtles make similarly long journeys to their nesting beaches. Baltic eels travel some 5,000 km across the Atlantic to the Sargasso Sea (south of Bermuda) to spawn, their surviving offspring eventually swimming back to home waters. The world's record holder for long-distance flight on a regular basis is the arctic tern, which breeds in the Arctic, but migrates to the Antarctic for the winter (the southern summer). The bird's annual round trip totals some 36,000 km; no other animal on Earth enjoys more days of sun[7].

Yet, despite our awareness of this rich diversity, we know extraordinarily little about marine life. Ocean ecosystems occur on a much larger scale than on *terra firma*, and the linkages between them are enormously complex, with marine ecological boundaries flowing into each other in a seamless web. With much of the planet unexplored, the Earth's land-based biodiversity may very well shrink in comparison to the tremendous variety found in the world ocean.

Where land and sea meet – coastal regions

Coastal ecosystems are among the most biologically productive of the entire planet. Wetlands, such as estuaries, salt marshes, and mangrove swamps, produce more wildlife, in terms of sheer volume, and more primary plant growth than any other terrestrial habitat. They make efficient natural percolators, filtering out pollutants washed off the land. They also trap and stabilize sediments, modify wave action, act as important staging and stopover sites for migrating waterfowl and shorebirds, and provide vital spawning, feeding, and nursery areas for commercially important fish and shellfish[8]. The Chesapeake Bay, for example, covering just 5,700 km^2 on the east coast of the United States, produces half of the country's catch of blue crabs, 90 per cent of its softshell crabs, and 15 per cent of its oysters[9].

Mangroves cover roughly 182,000 km^2 of intertidal, lagoonal, and riverine land throughout the world[10]. They provide habitats for over 2,000 species of fish, shellfish, invertebrates, and epiphytic plants, as well as for marine mammals such as dugongs and manatees, and their forest canopies provide homes for hundreds of species of birds. Mangroves manufacture large quantities of nutrients which sustain a wealth of marine life, making the first link in a long food chain that sustains offshore fisheries.

Seagrasses, the only land plants that have returned to the sea, are found in a wide swathe around the world in both temperate and tropical shallow seas. Like mangroves, they trap and consolidate sediment, clarifying the water, and provide nurseries, shelter, and food for a host of marine life, including commercially important fish and shellfish. They too are extremely fertile, with productivity levels comparable to those of agricultural croplands. The crude protein levels found in tropical eelgrass and dugong grass, for example, reach as high as 23 per cent of dry weight, higher values than those for terrestrial forage grasses. Furthermore, seagrasses can grow as fast as cultivated corn, rice, or tall-grass prairies without requiring chemical fertilizers[11].

Coral reefs are biological wonders, among the largest and oldest living communities of plants and animals on the planet, having evolved some 200 million years ago. Covering 255,000 to 600,000 km^2 – less than 1 per cent of the Earth's surface[12] – they rival tropical rainforest in their biodiversity. Southeast Asian reefs alone support an estimated 5 to 15 times the number of species of fish found in the North Atlantic. The Great Barrier Reef of Australia shelters 400 species of coral providing habitat for more than 1,500 species of fish, 4,000 different kinds of molluscs, and 400 species of sponge[13]. Nearly a third of all fish species live on coral reefs, and between 70 and 90 per cent of all fish caught by coastal fishers in tropical Asia are reef-dependent at one time in their lives[14]. Reefs protect coastlines from storm damage and reduce beach erosion; they provide homes, breeding areas, nurseries, and food for upwards of 1 million species; and they form an important link in cycling nutrients from the land to the open ocean.

Rocky shores, found particularly along coasts where erosion is high, can also be rich productive habitats. In temperate seas, such as the Atlantic shores of Europe and along the Pacific coast of the United States, cold nutrient-rich water fosters a profusion of life, including extensive mats of algae and seaweed. Large brown algae, known as kelp, is the marine equivalent of old-

WWF-Canon/Tony Rath

growth forests. Growing up to 40 metres in length, its dense 'foliage' provides cover and living space for hundreds of species of fish, invertebrates, and marine mammals[15].

The species diversity of sandy shores and mudflats is much lower than that of other coastal ecosystems, but those species that do occur there do so in great numbers. The massive populations of worms, clams, snails, and crabs that inhabit them provide a vital food supply for hundreds of species of shorebirds and waterfowl[16]. The extensive mudflats of the Waddensee, in northern Europe, for instance, provide feeding and staging areas for some 12 million waterfowl and shorebirds.

Open oceans and deep seas

Though some areas of the open ocean, such as the South Pacific, are nutrient-starved and fish-poor, others are prolific. In areas such as the North Atlantic, where currents converge and send a surge of nutrients to the surface, marine life thrives. The polar seas, once thought to be marine deserts with their frigid storm-tossed waters, produce enormous quantities of marine biomass due to nutrient-rich water which flows towards the Arctic and Antarctic. In the waters of the Southern Ocean, annual production of phytoplankton amounts to over 600 million tonnes. This, in turn, supports large quantities of zooplankton, about half of which is composed of crustaceans known as krill, whose standing stock is estimated at 500 million tonnes. This constitutes the basis of the Antarctic food chain, providing the main fare for the region's fish, squid, birds, and marine mammals, and most notably the 16 species of whales that are found there[17].

WWF/Anna Zuckerman/EYE ON THE WORLD

Oceanographers and marine biologists are learning that the abyssal depths where no light penetrates teem with many and bizarre forms of life. About 50 per cent of the planet is covered by ocean over 3,000 metres deep which may harbour millions of species. Entire colonies of animals new to science, including metre-long worms, large mussels and even crabs, have been discovered living around deep-sea hydrothermal vents that spout a hot, nutrient-rich broth of minerals into the water. So hot are these vents that the water around them nearly reaches boiling point. Here the primary producers are bacteria not dependent on sunlight. Many such highly specialized organisms await discovery. Only around one-tenth of the 290 million km^2 of the seabed has actually been explored and charted[18]. We know more about the moon than our own ocean world.

WWF-Canon/Jim Thorsell/IUCN

"The sea is the Palauans' highway, recreation area, and, above all, their paramount source of odoim or animal food... The chief traditional occupation of men is fishing, and there is no higher accolade than to be called a 'real fisherman'."

R. E. Johannes, 1981
Words of the Lagoon: fishing and marine lore in the Palau District of Micronesia

WWF-Canon/Meg Gawler

Some 14 per cent of the world's people draw their livelihoods, directly or indirectly, from the sea.

2. Sustenance and services from the sea

The world's seas have sustained and nurtured humanity for millennia, ever since early man emerged from the savannahs of East Africa some 2 million years ago. Ocean and coastal environments provide us with a cornucopia of products – everything from food, fibre, leather, and building materials to minerals, drugs, and medicines. Coasts are also the primary destination for millions of tourists every year, attracted to the sea by the endless recreational possibilities – swimming, snorkelling, diving, water sports, boating, sailing, fishing, and wildlife viewing amongst them.

It has been calculated that the world's ecosystems provide goods and services worth at least US$33 trillion annually, of which 63 per cent (US$20.9 trillion) is contributed by the oceans. Over half (US$12.6 trillion) of the oceans' contribution is from coastal ecosystems[19]. The haul of seafood alone is valued at around US$70 billion a year, giving direct employment to some 200 million small-scale and commercial fishers[20]. Perhaps as many as 500 million people draw their livelihoods indirectly from the sea: processors, packers, shippers, and distributors of seafood; shipbuilders and outfitters; those working in marine-based tourism, including cruise ships, tours, resorts, and dive shops; and the recreational fishing industry, amongst others[21].

■ Fisheries

As protein providers, the oceans are unequalled. Between 1991 and 1995 commercial fleets landed on average 84 million tonnes of seafood a year, by far the largest catch of a wild food source[22]. As a global average, seafood provides close to 20 per cent of the world's total animal protein intake[23]. In some regions, such as Southeast Asia and the South Pacific, the sea's bounty provides up to 90 per cent of all animal protein in daily diets. Some 950 million people, many of them desperately poor, rely on the sea as a major source of nutrition[24].

Most of the commercial take is landed by fishing fleets from just six countries: China, Peru, Chile, Japan, the United States, and Russia. In recent years, as the more valuable demersal (bottom-living) species have been overfished, there has been a switch to the less valuable pelagic (oceanic) species, such as Peruvian anchoveta, South American pilchard, Japanese pilchard, Chilean jack mackerel, and Alaskan pollack. The pelagic fish catch increased from around 6 million tonnes in 1950 to 44 million in 1994, and now makes up about half of the total annual take of seafood. Catches of four of the most valuable demersal species – silver hake, haddock, Cape hake, and Atlantic cod – have decreased by close to 70 per cent over a similar period[25].

WWF-Canon/Tony Rath

It has been estimated that 80 to 90 per cent of all commercial fish are caught within 320 kilometres (km) of shore[26]. More than half of all commercial fisheries production in the entire Pacific Basin, for instance, is concentrated along the coasts of Chile and Peru, where upwellings of deeper, nutrient-laden water create one of the world's richest fishing areas[27]. In tropical waters fishers rarely have to venture further than 25 km offshore. One hectare of mangrove forest in the Philippines, if properly managed, could produce an estimated annual yield of 200 kilograms (kg) of molluscs, 100 kg of fish, 40 kg of sea cucumber, 25 kg of shrimp, and 15 kg of crabmeat, as well as providing a nursery area for up to 400 kg of fish and 75 kg of shrimp that mature elsewhere[28]. In Puget Sound, Washington, United States, it was found that, when all factors are taken into account (the energy derived from the system as well as the nutrition the seagrasses generate for oyster culture, commercial and sport fisheries, and waterfowl), a little over a third of a hectare of eelgrass has an annual value of more than US$400,000[29].

WWF-Canon/Tony Rath

Some 50 per cent of the fish consumed around the world today is caught by small and medium-scale fishers in inshore and coastal areas[30]. Although fish are the main source of protein for enormous numbers of coastal people, nearly half of the catch today is exported, much of it from the South to the North, compared to just one-third in 1980. Between 1970 and 1990, the value of fish traded increased more than twelvefold, and global export earnings from fish now exceed those of coffee, tea, cocoa, and sugar combined[31]. This has some clear implications for food security. Less than a quarter of the world's population consumes over one-third of the catch, with industrialized country consumers accounting for 85 per cent of world imports by value. Developing countries supply more than 70 per cent of the world's canned tuna and 77 per cent of frozen shrimp and prawns[32].

While world fish catches have levelled off or declined since 1989, aquaculture is booming. It now accounts for some 20 per cent of global fish production, having more than tripled from 6.7 million tonnes in 1984 to 23 million tonnes in 1996[33]. One-third of this comprises marine species, of which shrimp and prawns are by far the most valuable. Shrimp farming has grown more than sixfold in the past decade, and has become a major export industry in Thailand, Ecuador, India, Bangladesh, Indonesia, and China. Production in Thailand alone has risen from 900 to 277,000 tonnes since 1985. Fish farms employ about 9 million people worldwide, and are a predominant source of employment in some coastal areas such as northern Norway and Thailand. The Food and Agriculture Organization of the United Nations (FAO) estimates that, by 2010, aquaculture (inland and coastal) could supply around 39 million tonnes of fish – an increase of 70 per cent over current production[34]. Since most species involved in aquaculture need animal protein as food, such an increase would mean a huge growth in fishmeal production, and thus in the use of a protein source that could be made available more directly to humans.

Other living resources

Marine resources of value to humans come in many shapes and sizes, and some ecosystems, like coral reefs, provide multiple benefits. Most tropical fish destined for the aquarium trade are captured on coral reefs and in seagrass beds. The popularity of reef fish continues to grow; the trade in these species is estimated to be worth around

US$200 million a year[35]. Corals are also mined for their lime, which in many tropical countries is an essential ingredient for cement production. They are also collected and used live for the aquarium trade, or bleached and cleaned for the souvenir industry. In the early 1990s, Indonesia, the world's main supplier, was exporting about a million pieces of live coral annually to markets in Europe, Japan, and the United States. Demand for ornamental shells is also large, mainly for use as coffee-table ornaments, jewellery, buttons, and other forms of shellcraft. In the mid-1980s, the Philippines, the centre of the world shell trade, was exporting several thousand tonnes of shells, shell craft, and both raw and worked mother-of-pearl – the 'nacre' or thick iridescent layers found inside the shells of several mollusc species[36].

WWF-Canon/Meg Gawler

Many marine organisms are used in traditional medicines in Asia. Treatments made from seahorses, for example, are believed to benefit complaints ranging from respiratory disorders to sexual dysfunctions. China's economic growth since the mid-1980s has caused an immense surge in the demand for such products, largely taken by small-scale and subsistence fisheries[37]. One of the reasons that marine products are so prevalent in such medicines may be that they contain pharmaceutically active compounds. Some 500 marine species identified so far possess chemicals that have antimicrobial, anticoagulant, and cardioactive properties, or could help to fight various cancers. Such species may prove invaluable in developing powerful drugs and medicines. The Australian Institute of Marine Science has even isolated a compound that protects coral from sunburn, giving it great potential for application in new, more effective sunscreen products[38]. Increasingly, such products are worth millions to the pharmaceutical industry: one compound derived from a sea sponge and used to treat herpes is worth between US$50 million and US$100 million a year[39].

Marine and coastal plants are also extensively used. Along rocky coasts in temperate seas, around 1,500 species of brown algae, including giant kelp, are exploited for their natural compounds, such as algin, used as emulsifiers in various food products, drugs, textiles, paints, and paper. Kelps and other brown algae collected from wild and commercially farmed populations are worth around US$150 million a year[40]. Mangrove forests provide literally dozens of products – from wood, tannins (used as preservatives for wood, fishing poles, and nets), and glue to synthetic fibres, cooking oil, and dye for cloth[41].

Oil, gas, minerals, and other non-living resources

The oceans provide a huge range of other resources: energy in the form of oil, gas, and tidal power; building materials in the form of sand and gravel; salt, which is used for food and industrial purposes; and valuable minerals. For many of these, exploitation has been limited by human skills and technology but this is changing fast, and with these changes come increasing threats from pollution and habitat disturbance.

The world's continental shelves contain vast amounts of oil and gas, and exploration is now under way on the deeper slopes. Submarine reserves of these two fossil fuels represent at least one-third of the estimated global total, and as much as 85 per cent of such hydrocarbon reserves is found within 320 km of land. Technological advancements in the past three decades have spurred the exploration and development of offshore oil and gas fields. Currently, about one-quarter of all oil and gas produced every year is sucked out of the seabed[42].

Deep ocean basins also contain a wealth of minerals, offering great potential for seabed mining. Deposits consist mainly of polymetallic nodules (often referred to as manganese nodules) and metallic muds. The nodules are made up of mineral salts which crystallize from seawater and contain valuable metals such as manganese, iron, cobalt, chrome, vanadium, and titanium. Geologists estimate that the Pacific Basin alone may contain 40 billion tonnes of manganese, over 40 times the estimated continental reserve. Some 2,000 metres down in the Red Sea, between the Sudan and Saudi Arabia, geologists have discovered metallic muds containing extractable amounts of zinc, iron, copper, and lead[43].

Tourism

Tourism is the largest industry in the world and now sustains more than 10 per cent of jobs globally[44]. Because people are naturally drawn to coasts, increasingly as places to relax, coastal and marine-based tourism is one of the fastest growing sectors of the tourist industry. Half a billion people took international holidays in 1996, a figure that is expected to keep climbing in the years ahead[45]. Of these, three-quarters ended up in coastal or near-shore areas. The Mediterranean Sea attracts over 150 million tourists a year[46], and islands and coastal nations, such as Belize in the Caribbean, also host huge numbers[47].

In the future, marine-based and coastal tourism is expected to expand significantly, as more people have more disposable income, and new technology provides ever-increasing varieties of leisure activities. No longer do people simply want to sit and watch the ocean; they now want to be scuba-diving, in boats or on jet-skis. The Mediterranean alone is expected to have to cope with upwards of 350 million tourists a year by 2025[48]; coastal and marine-based tourism in Asia and the Pacific is similarly booming. Unlike land-based tourism, which focuses to a large extent on cultural sites, marine and coastal tourism is specifically 'environment' based, and needs clean beaches and water, un-spoiled landscapes, and wildlife to view.

WWF/Y Lefèvre/BIOS

Transport and navigation

Long before roads connected countries to each other, the world ocean was the principal route of colonization and commerce. Before aeroplanes reduced the size of the planet, the world's seas played a major role in the development and spread of cultures and ideas, religions, and political systems. Many of the world's foremost civilizations rose and fell according to their access to, and use of, the sea.

The oceans remain the great highways over which goods are transported, and there are tens of thousands of vessels at sea at any one time. More than 4 billion tonnes of cargo are shipped by sea every year, 80 per cent by volume of the total amount transported internationally. About 2 billion tonnes of this consist of petroleum products – crude oil and natural gas – the bulk of which is shipped out of the Arabian Gulf to ports in Asia, Europe, and North America[49].

WWF/Michel Gunther/BIOS

WWF/Michel Gunther/BIOS

"Man's fingerprint is found everywhere in the oceans."

UNEP
The State of the Marine Environment, 1990

The rapid growth of coastal cities and the ever-increasing popularity of marine tourism demand that urgent attention be paid to the health of our coastal ecosystems.

3. Plundering the blue planet

The oceans are essential for life on Earth, yet humanity continues to treat the blue planet in a cavalier and thoughtless fashion. The world ocean, once thought to be able to provide an inexhaustible supply of food and other products, has been pillaged in the name of progress. The sea's resources are not unlimited, nor 'free', as previously thought, and exploitation comes with an increasingly expensive price tag. We are learning, however slowly, that many marine ecosystems can be destroyed as fast as terrestrial ones.

Population and development

Since the end of the Second World War, and particularly during the recent globalization of the economy, coastal cities – long-time centres of commerce and trade – have grown rapidly in both population and economic clout, turning into financial hothouses, providing the bulk of new jobs, and energizing economies. Over half the gross national product (GNP) of many countries is generated in their coastal cities. In Latin America, nearly two-thirds of the 80 major cities are coastal, while in China around 100 million people have moved from rural land-locked provinces to the new economic zones on the coast since 1990. It is estimated that 60 per cent of the current world population lives within 100 kilometres (km) of a coastline, and by 2025 coastal areas could contain as much as 75 per cent of the global population – some 6 billion people[50].

The twin threats of unbridled economic development and overexploitation of coastal and marine resources have led to over half the world's coastlines suffering from severe development pressures. For example:

- the world has lost half its coastal wetlands this century. Over the last hundred years, about 7,500 km^2 of mangrove forests (or 4 per cent of the current total) have been destroyed or grossly degraded[51]
- seagrass beds are in retreat near virtually all inhabited coastal areas, and have been almost wiped out in many temperate seas, victims of coastal development, dredging, and pollution. For instance, over a 24-year period (1960-1984), seagrasses in the Chesapeake Bay on the east coast of the United States declined from more than 80,000 hectares to less than 14,000[52]
- 58 per cent of the world's coral reefs are potentially threatened by human activity, and some 10 per cent may be damaged beyond recovery; over 80 per cent of reefs in Southeast Asia, the most vulnerable area, are potentially threatened[53]
- close to 70 per cent of the world's beaches are eroding at rates greater than is natural because of human impact[54].

Fisheries in crisis

The loss of critical coastal ecosystems, combined with overfishing, has precipitated a collapse of commercial and artisanal fisheries in many seas. Some 60 per cent of the world's top 200 commercially important marine fish stocks are either overfished or fished to the limit[55]. Productivity in all but two of the world's 15 major fishing regions has fallen over the past few years. In the four hardest hit areas – the Northwest, Western Central, and Southeast Atlantic, and the Eastern Central Pacific – catches have plunged by more than 30 per cent since the peak year of 1989[56]. Species with a long life and low reproductive rate are particularly at risk[57]. Many stocks, like the bluefin tuna and swordfish, have become so rare in some areas that they are regarded as 'commercially extinct', with too few fish remaining to warrant the expense of trying to catch them.

"It is harder to kill off fish than mammals. But after 1000 years of hunting the Atlantic cod, we know that it can be done."

Mark Kurlansky, 1997
Cod: a biography of the fish that changed the world

Sharks are particularly threatened. Millions of sharks are taken every year for their meat, fins, hides, jaws, and internal organs to satisfy demand for sharkfin soup, medicines, and other products. Estimated world catches (including small species such as dogfish) rose from just over 600,000 tonnes in 1985 to over 700,000 tonnes in 1994[58]. Atlantic populations of sailfish, blue and white marlin, and swordfish have declined by 60 to 90 per cent since 1980 and the trend is continuing[59]. In some cases, as in the North Sea, the Northeast and Northwest Atlantic, Southeast Asia, and the Black Sea, the demise of fisheries has been nothing less than catastrophic for local economies. In maritime Canada, the collapse of northern cod populations in the early 1990s due to chronic overfishing threw more than 30,000 people out of work in Newfoundland. In the Black Sea, commercial fish catches dropped from 1 million tonnes in 1982 to 100,000 tonnes by 1992, a tenfold drop in ten years[60].

In addition to having a direct impact on target species, causing their decline and ultimate collapse, fishing has various other impacts. The numbers of top predators such as billfish are so low that fishers, both commercial and recreational, are now taking mostly juveniles before they have a chance to spawn. For many species, populations are being affected as their age, size, and genetic diversity are reduced through selective fishing. We are now fishing 'down the food chain': having long since depleted top predators such as tunas and sharks, and groundfish like cod, fishers are increasingly catching smaller fish species that feed further and further down the marine food chain. Not only does this fundamentally alter the diversity of marine ecosystems, but scientists fear that soon we may have little left to catch other than invertebrates such as jellyfish, krill, and zooplankton[61].

Many factors, as described below, have contributed to the fisheries crisis, a situation made worse by the fact that fishery scientists still have only a dim understanding of sustainable stocking levels and species interactions.

Overcapacity and harmful subsidies

According to official figures, the capacity of the world's fishing fleet increased by 4.6 per cent a year between 1970 and 1990, twice as rapidly as catches. However, a recent study commissioned by WWF and IUCN, which takes into account advances in technological efficiency as well as growth in the size of the fleet, estimates that the world's fishing capacity may have increased by as much as 155 per cent since 1970[62]. Much of this growth has been financed by huge government subsidies which have ensured an investment in fishing far beyond sustainable levels. The annual world catch is estimated to be worth US$70 billion (and perhaps as much as US$100 billion), but to obtain this an estimated US$22 billion is spent on subsidies[63].

In addition to getting bigger, fishing fleets have become more sophisticated at chasing down and catching remaining fish stocks. The technical capacity – or ability to catch fish – of world fleets increased fourfold between 1965 and 1995, far outpacing the growth in fleet size, with the result that fleets are now equipped with a whole armada of sophisticated technological gadgets. Fishers off the coast of New England in search of the highly prized and elusive bluefin tuna hire spotting planes to help them locate their prey, and some pelagic fleets even use satellite imagery to locate fish schools[64].

Another sign of steeply declining stocks is the increase in the number of 'fish wars' and related conflicts on the high seas. In 1995, Canadian officials ignited a transatlantic turbot war by seizing a Spanish trawler caught taking undersized fish just outside Canada's 200-mile limit. Conflict occurs not only between countries, but also within countries among commercial, recreational, and artisanal fishers. Gunfire is a regular feature of local fisheries disputes in South and Southeast Asia. In June 1993, artisanal fishers from a port near Trivandrum, in the Indian state of Kerala, torched 14 trawlers, seized four others, and held one trawler captain hostage. Although this stand-off ended in a stalemate, India has now banned foreign trawlers from its waters[65].

WWF-Canon/Michel Gunther

Bycatch

Fishers today catch not only the fish that they are targeting, but quantities of 'bycatch' – species other than those being targeted, and juveniles of the target species – which are hauled in along with the target catch. Most fishing vessels are not equipped to handle bycatch, even if it has commercial value, because it cannot be kept when there are quota restrictions on the species concerned, and the bycatch is simply tossed overboard, dead or dying. Worldwide, fishing fleets generally discard between 18 and 40 million tonnes of unwanted fish, shellfish, and other marine creatures every year. (The Food and Agriculture Organization of the United Nations recently revised this average to 20 million tonnes per annum based on a recalculation of currently available discard data, but this figure remains unverified.)[66]

Trawl nets drag in many more fish and other marine organisms than the target shrimp, squid, groundfish, and other profitable species. Well over three-quarters of a shrimp trawl catch may consist of non-target species – sea turtles, birds, marine mammals, sharks, skates, rays, crabs, invertebrates, and even commercially valuable fish. In the Sri Lankan shrimp trawl fishery the ratio of bycatch to target species can reach 11 to 1[67]. Shrimp trawls in the Gulf of Mexico drown some 55,000 sea turtles annually[68], and hundreds of thousands of seabirds, including threatened albatrosses, are killed in long-line fisheries each year[69].

Destructive fishing methods

Destructive fishing methods cause as much damage as overfishing itself in many parts of the world, destroying the habitats on which the fisheries depend. Artisanal and small-scale fishers who seek to feed their families are forced by declining resources into a destructive cycle of dynamite, poisons, and fine-mesh nets. Such fishing methods are often the last resort for marginalized people in coastal countries who have been driven to fishing as the only source of protein and income.

Cyanide fishing is now widespread in Southeast Asia, the Indian Ocean, and parts of the South Pacific. While capturing fish for the aquarium trade or to supply the restaurants of Hong Kong, China, Japan, and Taiwan with live reef fish, fishers dump hundreds of tonnes of cyanide on reefs every year. The trade is thought to be worth US$1 billion a year and is growing in value as more people become involved. Large species, such as coral trout, groupers, and humphead wrasse, sell for over US$120 a kilogram (kg) in Hong Kong restaurants[70].

Dynamite fishing is equally destructive and is widely used in South and Southeast Asia, the Pacific, East Africa, and even the Mediterranean. A bomb the size of a beer bottle can pulverize coral over an area 3 metres in diameter, destroying in seconds communities that may have thrived for decades. The profits may be quick, but the destructive effects of these techniques are long lasting, particularly on reefs. Corals grow slowly; brain corals at a rate of some 5 to 25 millimetres a year, so that one colony may take a quarter of a century to grow as big as a person's head[71].

It is not only small-scale fishers that cause damage. In many parts of the Barents and North Seas, trawlers scour the bottom at least once a year, destroying entire ecosystems in their quest for the more valuable bottom-dwelling fish. Where trawlers operate in vulnerable habitats such as seagrass beds, immense damage can result, comparable to clear-cutting in forests[72].

Farming the sea – a solution creating problems

As wild fisheries have collapsed, commercial interests have led to the proliferation of aquaculture operations. In order to make room for extensive pond culture, millions of hectares of nature's free fish farms – mangrove swamps, salt marshes, and mudflats – have been converted into artificial environments. For example, Ecuador has lost 14 per cent of its mangroves and 40,000 hectares of salt marshes to shrimp farming; Thailand has lost half its mangrove forests to this industry[73]. This trend has also contributed to the further impoverishment of capture fisheries by depriving fish and shellfish of vital spawning, nursery, and feeding areas. This is of very real concern in poor coastal communities where fishing is the main livelihood. The replacement of mangroves with shrimp ponds can have a particularly negative effect on shrimp populations themselves, since shrimp are dependent on mangroves at critical stages in their life cycles.

Most large aquaculture operations offer little employment to local residents, other than low-paid menial labour. The majority of the profits – over US$36 billion in 1995 – go to large conglomerates or rich entrepreneurs, and the produce is largely exported to the developed world, contributing to a 'protein drain' from South to North[74].

Another aspect of fish farming tends to be overlooked by its proponents. All pond or cage-raised fish require food, increasingly provided as highly nutritious pellets made from small pelagic fishes such as anchovies, sand eels, and capelin. It takes, for example, about 5 kg of ocean fish reduced to fishmeal to raise 1 kg of farmed salmon or shrimp[75]. It has been estimated that, in 1990, about 25 per cent of the primary productivity of the North Sea was diverted into food to produce 130,000 tonnes of farmed salmon in Norway[76]. Shrimp and salmon farming are now considered two of the most resource-intensive methods of food production in the world[77]. Such systems increase, rather than decrease, the pressure on wild fish stocks.

Bad housekeeping practices by many fish farmers have also contributed to the pollution of coastal and near-shore waters, with some fish-farming operations producing up to a tonne of waste for every tonne of fish raised. Exotic species under cultivation often escape and become established in local waters, where they compete with indigenous species. Of all the exotic aquatic species that occur in Europe, 30 per cent originated in fish farms. Non-native Atlantic salmon, which have escaped from offshore pens in the US Pacific Northwest, now compete with endangered native salmon species, frustrating government efforts to restore wild populations[78].

Our fouled seas

Rapidly expanding coastal populations, the explosion of coastal cities and ports, rampant industrial growth, intensification of agriculture, extensive deforestation, and dredging have all contributed to the pollution that affects so many of the world's seas. The oceans are the ultimate sink for all pollution, 70 to 80 per cent of which originates from land-based sources, including atmospheric emissions[79]. Globally, some 450 km^3 of wastewater – from untreated or partially treated sewage, industrial effluents, and agricultural runoff – are carried into coastal waters by rivers and streams every year[80]. Coastal urban areas often dump raw wastes directly into the sea. Other sources of pollution include exploration of the seabed for oil and minerals, dumping of industrial and municipal wastes at sea, and the routine discharge of oily engine wastes and bilge slops from ships. Every year, 30 to 50 million tonnes of untreated or partially treated sewage are dumped in the Mediterranean Sea[81], while 50 to 60 million tonnes of untreated (or partially treated) municipal wastes flow daily into the Yellow Sea from China[82].

The main contaminants are toxic chemicals, hydrocarbons and oil, heavy metals (such as mercury, lead, cadmium, and arsenic), organic matter, nutrients, and sediments (such as nitrates and human and animal wastes), marine debris and solid waste (from non-biodegradable plastics to decommissioned oil and gas rigs), and radionuclides (from nuclear power plants and reprocessing facilities, the dumping of nuclear wastes, and nuclear testing).

Toxic chemicals

Every sea and every continent, from the tropics to the once pristine polar regions, is contaminated by toxic chemicals. Each year, thousands of tonnes of synthetic chemicals, including pesticides, are produced and over 1,000 new chemical compounds enter the market. The oceans are the final 'sink' for such toxic chemicals, which are found in pesticides, consumer products, industrial contaminants, and numerous other sources. They have been found in marine animals from the Arctic to the Antarctic. Many are equally hazardous to humans, having either immediate acute impact, or slower cumulative effects, as with carcinogens[83].

WWF/K H Sandrock Hoppe

"Monterey [California] as a collecting ground is already greatly injured, and will probably be nearly ruined before long, on account of the Hotel del Monte, the new town of Pacific Grove and the increased population of old Monterey, all the sewage of which is turned into the bay in front of the town. Beaches which formerly would afford several hundred species are now nearly bare, or offensive with stinking black mud. Old collectors will learn this with regret."

Dall, W.H., 1892
Letter to Editor
Nautilus 6: 48

Synthetic chemicals that degrade slowly or not at all, such as the pesticides aldrin and dieldrin, are known as persistent organic pollutants (POPs). These are particularly dangerous because they remain in the environment for long periods of time, gradually bioaccumulating up the marine food chain. Top predators, such as polar bears, whales, dolphins, and eagles, are often at risk. White-tailed sea eagles in the Baltic, for instance, accumulated such high concentrations of polychlorinated biphenyls (PCBs) in their bodies that they laid soft-shelled eggs which failed to hatch[84].

POPs and several other synthetic chemicals have become known collectively as endocrine disruptors, because they mimic hormones which play a key role in sexual differentiation and brain development. They can thus have particularly deadly, long-term effects on both human and wildlife populations. Even in small doses they can cause neurological damage and reproductive abnormalities, and disrupt the development of the reproductive, immune, nervous, and endocrine systems. Over 50 chemical compounds, many in use for decades, have been identified as endocrine disruptors, including PCBs, dichlorodiphenyltrichloroethane (DDT), aldrin, dioxins, and tributyltin (TBT). Most of them are found in pesticides and plastics, while others are released during industrial processes[85]. TBT, which is used in anti-fouling paint on ships' hulls and other underwater structures, causes 'imposex' – the development of male genitalia on top of female genitalia – in whelks and other marine molluscs. It also caused the decline in production of native oysters in the 1970s and 1980s in northwest Europe and, like most other endocrine disruptors, is passed up the food chain to seals, dolphins, whales, and possibly humans.

When seal populations in the Baltic and North Seas were decimated by major outbreaks of disease in the 1980s, many of the corpses examined were found to contain high levels of PCBs and other endocrine disruptors. High levels of industrial chemicals have been linked to reproductive disorders and immune system failures[86]. Whales in all of the world's oceans now carry enormous toxic burdens[87] and are often plagued with tumours, reproductive problems, and heavy metal poisoning. Endangered beluga whales in the mouth of Canada's St Lawrence River have such high levels of PCBs in their blubber that under Canadian law they qualify as 'toxic waste dumps'[88]. Albatrosses nesting on the remote Midway Island in the Northern Pacific and polar bears on Kongsoya Island, high above the Arctic Circle in Norway's Svalbard archipelago, are also carrying hazardous levels of chemicals, including PCBs, dioxins, and DDT[89].

Non-toxic pollution

Every year river systems carry roughly 25 billion tonnes of soil eroded from the land into estuaries and near-shore waters. Much of this is due to human activity such as agriculture and forestry, often carried out unsustainably. Increasingly, abnormal conditions are created, causing turbidity and suffocating life on the sea or estuary beds by smothering benthic communities and clogging up animals that filter-feed. In Queensland, Australia, rivers now carry an estimated three to five times more sediment to coastal waters than before European settlement[90]. Dredged sediment from harbours and waterways, combustion processes, and colliery wastes further increase the load.

Nitrates and phosphates contained in fertilizers, nutrients from untreated sewage, and other organic pollution from various sources, including those listed above, make their way into coastal waters and overfertilize such ecosystems. Seagrasses and coral reefs are particularly susceptible to nutrient pollution since they depend on clear, clean water for photosynthesis and maintenance. As nutrients increase, not only does the water become more turbid with microscopic algae, but large bottom-living algae grow faster and eventually outcompete the corals.

But all ecosystems suffer. During the first half of the 1990s, 1.2 million tonnes of nitrogen and 800,000 tonnes of phosphorus were discharged into the Baltic Sea, mostly from sewage and agricultural chemicals[91]. In the Black Sea all waters below 150 to 200 metres are anoxic; only 73,000 km^3 of near-surface water (10 per cent of the total volume) have enough oxygen to sustain life higher than microorganisms. In both these semi-enclosed seas, the natural processes that turn deeper waters eutrophic have been aggravated by the massive amounts of urban, agricultural, and industrial pollution flushed into them over the past four decades[92].

WWF-Canon/Paul Glendell

Large influxes of nutrients can trigger massive algal blooms that snuff out marine life. When the algae die and decay, they rob the water column of oxygen. Evidence suggests that harmful algal blooms, or red tides, are on the increase throughout nearly every sea on the planet, possibly propelled by climate change. One massive algal bloom that ravaged the Baltic and North Seas in 1988 was so destructive of marine life that it became known as the 'marine Chernobyl'. Covering some 75,000 km^2, it wiped out hundreds of thousands of invertebrates and fish[93].

Marine-based pollution

The oceans not only have to contend with land-based pollutants, they are also under assault from sea-based pollution – from oil and chemical spills, hydrocarbon extraction, routine shipping, and deliberate dumping of wastes at sea.

Of all these sources of marine pollution, the discharge of oily engine wastes and bilge slops from routine shipping operations is perhaps the worst because it is ubiquitous and chronic. Despite international conventions and laws, ships discharge from 5 to 50 million tonnes of oil at sea every year[94]. Oil spills, which dump massive amounts of oil into marine and coastal waters, can cause immediate and in some cases long-lasting damage to ecosystems and marine life. The ill-fated oil tanker *Exxon Valdez*, which struck a reef in Alaska's Prince William Sound in March 1989, dumped 257,000 barrels of crude oil into near-shore waters. In the aftermath, some 5,000 sea otters perished, along with 300,000 seabirds and 300 bald eagles. By 1995, ecosystems had still not recovered from the toxic shock and its longer-term effects[95]. Since then, there have been three major oil spills in northwest Europe alone (the *Aegean Sea* spilling 80,000 tonnes, the *Braer* 84,500 tonnes, and the *Sea Empress* 72,000 tonnes) as well as notable spills in the Mediterranean, and off the coasts of Australia, Indonesia, Japan, Malaysia, and South Africa.

The routine discharge of oiled drilling muds, production water, and chemicals from offshore oil and gas extraction has also caused serious pollution in and around drilling platforms. Estimates in 1991 from the North Sea suggested that around 23,000 tonnes of oil, 84,097 tonnes of drilling chemicals, and 5,934 tonnes of production chemicals were being released this way into the marine environment annually[96]. Hydrocarbon exploration and exploitation is expected to increase, particularly in the Arctic and off Australia, the Falkland Islands, the UK continental shelf, and Vietnam, among other places.

Seabed mining is rapidly becoming technologically feasible and cost-effective: an Australian/ Papua New Guinea company is slated to begin mining operations off the coast of Papua New Guinea at depths of 1.2 to 1.7 km. This form of resource use threatens to introduce as yet unknown sources of pollution to the marine environment.

Until the mid-1990s, countries of the Organisation for Economic Cooperation and Development (OECD) also dumped millions of tonnes of sewage sludge, dredgings, and certain kinds of industrial waste at sea. In recent years, most ocean dumping has been banned, but it continues nonetheless, even where declared illegal[97].

Solid waste and debris

There is a more visible form of pollution that, at least in the short term, is just as deadly as invisible chemicals: the proliferation of solid wastes. An estimated 6 million tonnes of rubbish enter the sea every year[98]. Marine debris is now found on the remotest Pacific islands and in the deepest ocean trenches. The main sources are the packaging industry, urban and domestic rubbish, abandoned gear from fishing boats, and rubbish disposal by shipping. Plastics, in particular discarded fishing nets and polypropylene packing pellets, wreak havoc with marine life.

Alien species

Pollution can take many sinister forms. One of the more insidious is the introduction of exotic or non-native species into marine environments. These marine invaders are on the increase in nearly every sea for a variety of reasons, of which the most important is probably the big increase in international shipping traffic. Many exotics, from micro-organisms to fish, hitch-hike in the ballast waters and cargoes of ships. Until the beginning of this century, rock or sand was usually used as ballast, but the change to water opened the way for major invasions. Other aliens travel in other parts of ships, on anchor chains, or as fouling organisms attached to hulls, while yet others are escapees from research laboratories and aquaria. Many exotics are robust opportunists, able to take advantage of impaired water quality and impoverished habitats, displacing native communities of plants and animals in the process. Some 60 introduced species have been found in the Baltic Sea; over 150 have been identified in Australian waters and at least 8 of these are major pests, including a Japanese seaweed, a European crab, and a Pacific sea star. The rate of invasion into the San Francisco Bay, which was about 1 species a year up to the 1960s, has increased to 1 species every three to four months in the last 30 years.

WWF-Canon/Mauri Rautkari

No matter where they come from, alien species frequently disrupt the ecosystems which they invade. For example, in some parts of the Black Sea, 90 per cent of the marine fauna consists of the comb-jelly *Mnemiopsis leidyi*, which feeds on the eggs and larvae of many commercially valuable species of fish. The comb-jelly was imported inadvertently from North America in the ballast water of a merchant ship, and has contributed to the collapse of several fisheries and the biological impoverishment of the entire sea. In the 1980s a Chinese clam was introduced to San Francisco Bay and soon dominated the ecosystem, using up the greater portion of the Bay's phytoplankton and depriving other, native species of habitat and food. Similarly, the green algae *Caulerpa taxifolia*, common in many warm seas, has spread relentlessly through the Mediterranean since 1984, when it was first observed near the well-known Monaco aquarium, from which it had almost certainly escaped. It is now competing with the native *Posidonia* beds and posing a threat to many endemic algae[99].

Climate change – rising seas and warming waters

There is no longer any doubt that the world's climate is changing and this will have a number of impacts on the oceans. Sea levels are rising, sea temperatures are increasing, and there is potential for major changes in the oceans' largely temperature-driven current systems.

As the world warms, the oceans expand and rise slowly, incrementally. Sea levels are 10 to 25 centimetres higher than a century ago, and are expected to continue to rise for many years, with a predicted rise of 15 to 95 centimetres by the year 2100[100]. If the ice sheets of the Arctic and Antarctic begin to melt, then levels could rise suddenly and catastrophically. Increases of 1 metre or more would mean that large areas of low-lying island nations, like the Maldives and the Marshall Islands (most of which have maximum elevations of 2 metres or less), would disappear off the map. Many sections of continental coastline could also be inundated; a 1-metre rise would mean that 17 per cent of Bangladesh would be under water, and half the world's coastal wetlands of international importance (i.e. those listed under the Ramsar Convention) would be at risk[101].

Scientists are also concerned that changes in sea temperature could result in a radical change in deep-ocean circulation and thus patterns of upwellings of nutrient-rich water. It is known from El Niño studies that even small shifts in oceanographic and meteorological patterns can have enormous consequences for marine productivity and food-web stability and thus for commercial fisheries. Some fisheries might boom, while others would probably decline or move polewards. Several commercial stocks in the North Pacific, off California, including sardines, anchovy, and hake, declined in the mid-1990s, along with the population of sooty shearwaters. These events have been linked with a decline in zooplankton due to a 0.8°C rise in sea temperature since the 1950s[102]. High-energy climatic phenomena such as hurricanes and typhoons, and violent shifts in weather patterns are also predicted. Climatologists point to the increase in frequency and intensity of El Niño events over the past two decades as early warning signs of impending climate change. The 1997/1998 El Niño, which brought heavy rainfall to the west coast of North and South America and devastating drought to Southeast Asia, has caused billions of dollars of damage and made thousands homeless[103].

High temperatures also cause major damage to coral reefs, probably the ecosystem most sensitive to global warming. A temperature rise of just 1°C above normal can cause corals to expel their symbiotic algae (which provide them with food and give them their bright colours), a process known as coral bleaching. El Niño events since the 1980s have caused coral bleaching around the world, with the 1997/1998 El Niño causing the worst bleaching on record, and affecting reefs in all oceans. Although corals can regain their vital algal symbionts, many reefs have made a poor recovery, even from bleaching that took place 20 years ago. With the predicted increase in sea surface temperatures, we can assume that bleaching will become a more frequent phenomenon and put reefs at increasing risk[104].

"The future historians of science may well find that a crisis that was upon us at the end of the 20th century was the extinction of the systematist, the extinction of the naturalist, the extinction of the biogeographer – those who would tell the tales of the potential demise of global marine diversity."

Carlton, J.T., 1993
Neoextinctions of marine invertebrates *Amer.Zool.* 33: 499-509

Climate change, in combination with overfishing and habitat loss, will exacerbate the current perilous state of many marine species. Those dependent on plankton and other food resources affected by changing circulation patterns and increased sea temperatures will be especially at risk. Whale species could be particularly affected, as some of the greatest oceanographic effects may be felt in the polar seas, major feeding areas for many baleen whales[105].

Disappearing species

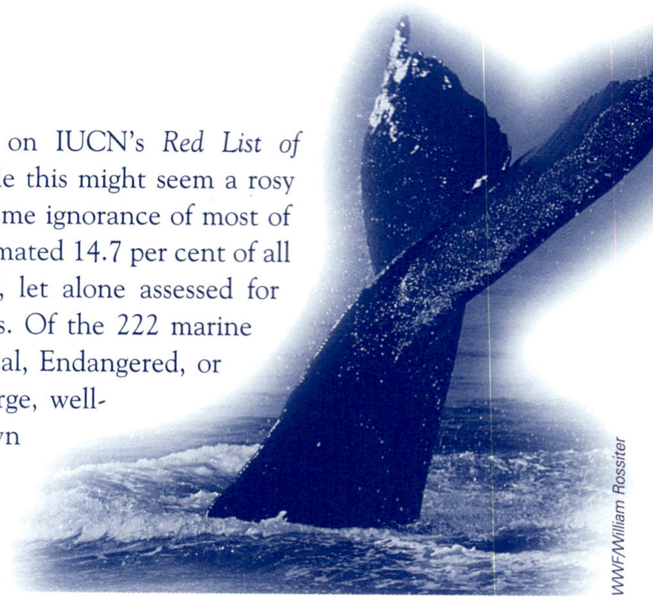
WWF/William Rossiter

Only 4.2 per cent of the species on IUCN's *Red List of Threatened Animals* are marine. While this might seem a rosy picture, in reality it reflects our extreme ignorance of most of the ocean's inhabitants – only an estimated 14.7 per cent of all marine species have been described, let alone assessed for population size, distribution or status. Of the 222 marine species listed in the categories Critical, Endangered, or Vulnerable, some 40 per cent are large, well-studied, and relatively well-known mammals and birds (see table). The status of the more than 14,000 known species of marine fish is much less well understood. Only a small proportion has been assessed and only 111 have been listed, and these as recently as 1996. Of the thousands, and possibly millions, of invertebrates, only 12 are included, reflecting more the level of our understanding than the threat to the species[106].

Marine species are affected by the whole range of factors described earlier in this chapter: habitat loss which is a major threat to, for example, many seals and turtles; bycatch as a result of long-line fishing, trawling, and other non-selective fishing gear (a major threat to albatrosses and turtles); mounting pollution (especially deadly for marine mammals such as whales and dolphins); and competition with alien species. Many species are subjected to a range of impacts. Marine turtles, for example, are threatened at all stages in their life cycles: their nesting beaches are taken over for tourist resorts, housing, and industry; subsistence fishers kill them for their meat and shells; others make a living collecting their eggs; and tens of thousands are drowned in fishing nets and shrimp trawls.

Exploitation, either for subsistence use or for commercial trade, continues to be a major threat for many marine species despite recognition of the need for conservation. After 30 years of working to conserve the world's whales, many populations have shown little sign of recovery from five centuries of indiscriminate slaughter and several are on the *Red List*. The population of the blue whale, which once numbered around 275,000, is now probably under 5,000. There are thought to be fewer than 1,000 northern right whales left and only a few thousand pygmy blue whales[107]. Equally striking is the fact that commercial fish species are now meeting the criteria for the *Red List*. The southern bluefin tuna, for example, is listed as critically endangered, having undergone a population reduction of 80 per cent over three generations[108].

There is also increasing conflict between commercial fishing and the food requirements of certain threatened species. Where fish species are being overfished, their predators may be at even greater risk as a result of a reduced food supply. For example, in the Southern Ocean, overexploitation of fish, krill, and squid could have serious consequences for many of the seabirds, seals, and whales that prey on these populations. In addition, fishers frequently view predators' feeding activities as detrimental to fish stocks and perceive them as a major economic threat, demanding culling programmes or themselves turning to either legal or illegal hunting. Seals, such as the endangered Mediterranean monk seal, have been most persecuted for this reason, but a number of cetacean species have also been affected[109].

The status of migratory species is particularly difficult to determine, since they are not connected to a specific habitat and range across thousands of kilometres of ocean. Tuna, billfish, and sharks, for example, are all highly migratory, and areas in the vast ocean where these species mate and spawn – such as the Gulf of Mexico for the bluefin tuna – are critical for their conservation. But for many species information is lacking. We simply do not know enough about the lives of these fish, their predator-prey interactions, or the environments through which they travel, to make intelligent judgements on how to conserve them[110].

In addition, there is growing concern about marine species which have little or no commercial or cultural value, but which have restricted ranges and could be suffering from damage to their habitats as more valuable species are overexploited and removed from the ecosystem. For example, recent interest in oil and gas exploration in the Northeast Atlantic has led to concern that cold-water corals such as *Lophelia pertusa*, which are barely known to science because of their deep habitat, could be at risk. Coral reef species may be especially threatened, particularly the numerous relatively sedentary invertebrates that have yet to be properly recorded. Given current trends in coral reef decline and existing knowledge of the diversity of coral reef species, it has been estimated that 2.6 per cent of reef species are already extinct and that over 10 per cent may disappear in the next 20 years. Such theoretical extrapolations have yet to be proven, but in the marine environment there is a high likelihood of species extinction occurring without anyone noticing. This is particularly true as marine scientists trained in taxonomy and biogeography, with the skills to identify and describe the myriad life forms of the sea, are themselves in ever-shorter supply[111].

Marine species on the IUCN *Red List and the CITES Appendices+**

	Red List	CITES Appendix I	Appendix II
Whales, dolphins	13**	22	All cetaceans
Marine otters	1	1 geog.pop.	1
Seals, sea lions	12	3	9
Sirenians	3	3	1
Birds++	61	4	1
Reptiles	9***	7**	1
Fishes+++	111	2	
Mollusca	10		10
Corals and coral-related species	2		All stony and black corals

NOTES:
* Categories Critically Endangered, Endangered, and Vulnerable only.
\+ Appendix I: Species which are at risk of extinction and in which international trade is prohibited.
Appendix II: Species which are vulnerable to exploitation but not yet at risk of extinction, and in which international commercial trade is permitted in a regulated manner.
++ Bird totals exclude waders.
+++ Fish totals exclude sturgeon.
** For some species, only certain populations are listed.
*** Totals include all 7 species of sea turtles.

"What we must do is encourage a sea change in attitude, one that acknowledges that we are a part of the living world, not apart from it."

Sylvia Earle, 1996
Sea Change: a message of the oceans

WWF-Canon/Hartmut Jungius

Community empowerment, social and economic incentives, education, and legislation based on sound science must combine to ensure an equitable and sustainable exploitation of marine resources.

4. Solutions for the ocean planet

Conservation efforts for oceans and coasts have in general lagged behind those for the terrestrial environment. There are a number of reasons for this, related to the characteristics of the marine environment described earlier. Certain overriding principles, which take these into consideration, must therefore be followed in developing effective management strategies.

- **All measures and practices to conserve marine biodiversity and ecological processes must take human needs into account.**

Given the significance of the oceans to humankind's social and economic well-being, measures to conserve marine biodiversity and ecological processes will only be successful if human needs are considered. As pressures on the land and its resources grow, the sea is increasingly relied upon for sustenance and recreation. Sustainable management of the oceans must become as high a priority as sustainable management of the land.

- **The concept of stewardship must be fostered through education and by increasing awareness at all levels.**

A broad public constituency demanding action is an essential component of sustainability in conservation. Large parts of the coastal and marine environment have traditionally been considered 'commons', accessible to all and thus highly vulnerable to overexploitation[112]. Unlike the concern for tropical rainforests – which are above water and highly visible – until recently there was no widespread public outcry to preserve the oceans and the numerous species they shelter, unseen except by marine scientists and recreational divers. Education and public awareness campaigns can reverse this, as is already being shown in some parts of the world: a 1997 poll in the United States, for example, found that 55 per cent of Americans believe that ocean exploration should take precedence over space exploration, and a large majority believes that the health of the oceans is in jeopardy[113].

- **Communities must be empowered to protect and manage their marine and coastal resources and local and national management capacity must be increased.**

Community management approaches in many parts of world – from traditional forms of resource management in the Pacific to the lobster fishery in Maine in the United States – are showing how effective empowerment can be achieved. Equally important is the need for a shift in the relationship between government and resource users. Collaborative or co-management – the sharing of management between government and stakeholders – is one solution, and is increasingly evident in marine resource management[114].

Community participation in Marovo Lagoon

WWF's community-based programme in Marovo Lagoon and adjacent islands in the Solomon Islands helps community groups to develop their own marine resource policies, setting aside sites for protection and recovery, controlling harvest of invertebrate and fish species, safeguarding local populations of endangered and *tabu* species such as dugong, turtle, and crocodile, and developing marine-oriented village tourism, including a small locally run hotel. The communal project sites will provide model marine conservation areas for the rest of the country.

- **Social and economic incentives for conservation and sustainable use must be created.**

One solution to current overexploitation may lie in humankind's growing dependence on the world's seas, offering the potential for social and economic incentives for their conservation and sustainable use. As on land, regulatory measures alone are unlikely to address the complexity of the issues, and the global market in marine products and services could provide additional incentives for more sustainable practices. For example, the World Bank has established a Marine Market Transformation Initiative (MMTI) to support partnerships that can provide sustainable economic alternatives to current practices in the fields of aquaculture, tourism, and the reef fish trade. Certification and ecolabelling, already having an impact in the forestry[115] and tourism sectors, are gaining increasing attention in the fisheries and aquaculture sectors (see below – 3.2).

Trade-related measures are another option. A vast international market drives the extraction of living marine resources; fishery products, valued at US$52 billion in 1995, make up a substantial portion of world trade[116]. Thus in theory, commercial trade rules could have an enormously positive influence on the sustainability of fishing practices worldwide. However, tensions between trade and conservation interests are increasingly evident, particularly in the fisheries sector. The World Trade Organization (WTO), for example, has ruled that US embargoes of canned tuna and frozen shrimp (from shrimp caught without turtle excluder devices or 'TEDs'), to protect dolphins and sea turtles respectively, are illegal under current trade rules[117]. This needs to be addressed urgently, and work must be intensified with the WTO to look at conflicts between trade rules and national policies in favour of conservation and other environmental issues.

- **The interconnectedness of the oceans must be recognized through appropriate cooperative transboundary and international mechanisms.**

Ocean waters and the organisms, nutrients, and pollutants that they carry do not respect national boundaries as they circulate around the planet. The activities of one nation may have an immediate impact on the coastal waters of other nations nearby and, over time, affect other waters and people thousands of kilometres away. A significant proportion of the world's seas lies outside the jurisdiction of any individual nation and is managed under the 'freedom of the seas' principle, regarded as freely available to all those with the means and technologies to exploit it. This, combined with the trend towards globalization of markets and economies, means that regional and international cooperation is vitally important[118]. Consequently, over 300 international treaties and programmes have been negotiated and adopted to deal with pollution, shipping, fisheries, mining, research, and other ocean issues.

The most important and comprehensive is the United Nations Convention on the Law of the Sea (UNCLOS), which entered into force in November 1994. This preserves many of the traditional freedoms of the high seas while increasing the authority and responsibility of coastal states for managing the resources within their 200 nautical mile Exclusive Economic Zones (EEZs)[119]. Over 100 nations have ratified UNCLOS, more have signed it, and many are indicating their intention to adhere to it. Most recent international agreements have been built on its widely accepted basic principles. Indeed, its obligations are implemented through other conventions and agreements specific either to particular regions, such as the pioneering Regional Seas Programmes of the United Nations

Environment Programme (UNEP), which have been running for two decades, and the 1991 Environmental Protocol to the Antarctic Treaty, which entered into force in January 1998, or those that deal with particular issues, such as fisheries and pollution (see below). In addition, there is a second comprehensive agreement of paramount importance for the protection and sustainable exploitation of the oceans: the Convention on Biological Diversity (CBD) (see box)[120].

The Convention on Biological Diversity (CBD)

The objectives of the CBD are the conservation of biological diversity, its sustainable use, and the equitable sharing of benefits derived from genetic resources. The Jakarta Mandate, adopted by the parties in 1995, is a declaration of intent that identifies five key issues where the CBD can play a specific role in ocean management: integrated coastal management, marine protected areas, sustainable use of marine living resources, aquaculture, and the introduction of alien species. WWF, IUCN, and the Centre for International Environmental Law have published a guide for governments on implementing the treaty in marine and coastal habitats[121]. This translates the general obligations of the CBD and the Jakarta Mandate into practical actions, many of which are based on successful case studies. These include the need to identify and monitor priority components of biodiversity, to build capacity to use and share the benefits of marine genetic resources, and to take responsibility for transboundary harm and global threats to marine biodiversity. The guide also highlights the links between the CBD and other relevant international instruments that must be fully implemented in order to harness the full strength of international law.

While many United Nations agencies are responsible for the protection of the marine environment and sustainable exploitation of its resources (such as the Food and Agriculture Organization (FAO), UNEP, the International Maritime Organization (IMO) and the Intergovernmental Oceanographic Commission (IOC)), no single agency is charged with an overview of oceans issues or is responsible for identifying and responding to emerging issues. The United Nations General Assembly has a mandate to address all of these, but in practice only dedicates a few days each year to a review of UNCLOS, and participation by non-governmental organizations (NGOs) and other oceans stakeholders is limited. The Commission on Sustainable Development (CSD) is convened every year to review implementation of Agenda 21, the principal outcome of the 1992 United Nations Conference on Environment and Development (UNCED or the Earth Summit), and in 1999 will focus on Chapter 17, which provides a comprehensive but general action plan for the world's oceans and coasts. Better integration is clearly needed between all these intergovernmental activities.

- **The precautionary approach must be applied and an ecosystem-based approach to management fostered.**

Current ignorance of many aspects of the oceans makes the adoption of the precautionary principle (that is – erring on the side of conservation rather than of exploitation and pollution) essential to the development of appropriate management strategies.

Furthermore, since oceans, seas, and coastal zones are ecologically linked across vast distances and are profoundly affected by inputs from land, freshwater, and the atmosphere, strategies must involve an integrated ecosystem approach. Marine biodiversity can be conserved effectively only if the habitats that sustain threatened species and fish stocks are maintained. Realization of this has led to the establishment of marine protected areas (MPAs), which are a good first step. However, impacts originating outside their boundaries, such as pollution, siltation, and overfishing, can make management efforts meaningless. To address this, integrated coastal management (ICM), which integrates the management of coastal areas with that of its watersheds, and of marine resources offshore with that of the high seas, is being introduced in many coastal countries. Widening the area that needs to be managed increases the challenge, however, and ICM is still in its infancy.

As understanding increases of how the precautionary principle and the ecosystem approach can be translated into on-the-ground management practices, their application should become easier, and current reluctance to adopt them in global and regional agreements concerning the oceans should decrease.

- **All solutions must be based on the best available science and knowledge, with encouragement given to the acquisition, dissemination, and exchange of information.**

As stressed earlier, human understanding of the oceans is still far from adequate. This, however, is no excuse for inaction. Decisions can be made on the basis of the information that is available, and further research and better data exchange and dissemination can be promoted.

Creating a sea change – a global marine policy

Reflecting the missions of both organizations (see box), WWF and IUCN have determined the following goals for this marine policy:

- to maintain the biodiversity and ecological processes of marine and coastal ecosystems
- to ensure that any use of marine resources is both sustainable and equitable
- to restore marine and coastal ecosystems where their functioning has been impaired.

WWF's mission is to conserve nature and ecological processes by:

- preserving genetic, species, and ecosystem diversity
- ensuring that the use of renewable natural resources is sustainable both now and in the longer term, for the benefit of all life on Earth
- promoting actions to reduce to a minimum pollution and the wasteful exploitation and consumption of resources and energy.

IUCN's mission is to influence, encourage, and assist societies throughout the world to:

- conserve the integrity and diversity of nature, and
- ensure that any use of natural resources is equitable and ecologically sustainable.

WWF and IUCN believe that significant progress can be achieved through five closely linked objectives relating to MPAs, threatened marine species, sustainable fisheries management, marine pollution, and ICM.

OBJECTIVE 1: The establishment and implementation of a comprehensive, global network of ecologically representative, well-managed MPAs designed to conserve areas of high biological importance and productivity.

MPAs are often thought of as modern concepts, but the idea of closing off areas of coastal waters to certain human activities flourished centuries ago in some regions, most notably the Pacific, where *tabu* areas were seasonally or permanently closed if, for instance, the weather was so bad that the normal fishing grounds could not be reached, or a certain food needed to be harvested for a village feast[122]. More recently, MPAs have been set aside under statutory legislation for a variety of reasons. Currently, there are over 1,000 of these areas, scattered like grains of sand around the world's seas. Though they can cover a considerable area, such as the 340,000 square kilometres (km^2) of the Great Barrier Reef Marine Park of Australia, most are fragmented and small, as little as 3 km^2, covering tiny slivers of coastal and near-shore waters. Apart from the 30 per cent of the oceans declared as whale sanctuaries by the International Whaling Commission (IWC), MPAs make up less than 1 per cent of the ocean's surface[123].

Well-managed MPAs contribute directly to the conservation of marine ecosystems and the prolific array of plants and animals they support, including threatened species. Protecting ecosystems such as estuaries, mangroves, seagrass beds, and coral reefs translates directly into more productive fisheries. MPAs in which there is a ban on fishing, or which have no-fishing zones (NFZs), enhance fish productivity and help restock and restore overexploited areas (see below – 3.1). And while reducing the impact of fishing, tourism, and other activities, local communities may be able to benefit economically from alternative livelihoods including carefully managed tourism, thereby taking pressure off overexploited marine resources. However, despite the many benefits of setting up and sustainably managing MPAs, their establishment is often met with considerable resistance.

Moreover, even when countries do set up MPAs, they often neglect to provide them with the resources necessary for good management. And even with adequate resources, MPAs are not without their difficulties: conflicts can arise between users who want access to the same sites for very different activities, such as fishers who want to remove fish and divers who want to watch them. Protected areas also tend to have a 'honey-pot' effect, attracting visitors with their promise of pristine habitats and diverse wildlife. Equally important, and even more difficult to tackle, are the activities taking place outside protected areas that cause pollution and siltation, and which can render management efforts meaningless. Potential solutions to some of these problems are identified in the activities below.

Priority activities

1.1 Assist in developing, strengthening, and implementing regional and global agreements and mechanisms for the establishment and management of MPAs, such as:

- the CBD, UNCLOS, the Convention on Wetlands of International Importance (Ramsar Convention), and the CSD
- the UNEP Regional Seas Programmes and other relevant regional agreements.

The CBD is the global agreement that currently has the most comprehensive role in relation to MPAs, requiring each country to establish a national protected area system. Through the Jakarta Mandate, nations are encouraged to take a systematic approach to the establishment of MPAs. Further consideration needs to be given to the role that UNCLOS might play, particularly in relation to MPAs on the high seas. The World Heritage Convention and the Ramsar Convention, although valuable tools for promoting protected areas that are, respectively, unique and outstanding, or coastal wetlands of international importance[124], need strengthening with respect to the marine environment, and new areas should be designated. The CSD also has a role to play, generating high-level political support for the coordinated implementation of these agreements. Several regional agreements, particularly in Europe, are now actively promoting the establishment of MPA systems, including the Convention for the Protection of the Mediterranean Sea against Pollution (the Barcelona Convention), the European Union Habitats Directive, the

Convention for the Protection of the Marine Environment of the North-East Atlantic (the OSPAR Convention), and the Helsinki Commission. These initiatives need further support, and similar efforts should be made in other fora such as the Cartagena Convention and the Nairobi Convention, the agreements drawn up under the UNEP Regional Seas Programmes for the Caribbean and East Africa respectively.

1.2 Ensure that this global network is truly representative by:

- **evaluating and identifying gaps, and creating new MPAs to address these, e.g. for cetaceans, pelagic species, and the seabed**

- **promoting the concept of MPAs in offshore, transboundary, and international waters and identifying priority sites and mechanisms by which they might be established and enforced.**

Many more MPAs are needed to ensure that all threatened species and their habitats, and all marine ecosystems and key ecological processes, are represented. The process of evaluating and identifying the gaps in the current network has been hampered by the lack of a globally accepted biogeographical classification. For example, although about 685 protected areas include mangrove habitat, these are very unevenly distributed in relation to the distribution of mangroves worldwide: there are 120 protected areas with mangroves in Australia, but none in Nigeria, which is one of the countries with the largest areal cover of mangrove forest in the world[125].

Efforts to 'fill the gaps' in the global network are nevertheless under way. IUCN, the World Bank, and the Great Barrier Reef Marine Park Authority have carried out a worldwide survey of MPAs and identified 155 existing and proposed MPAs as priorities for future action, reflecting the importance of their biodiversity. IUCN, with the National Oceanic and Atmospheric Administration (NOAA) and the IOC, is testing the use of Large Marine Ecosystems (LMEs) (large biogeographic units) as a basis for management. WWF is also developing a system for setting priorities based on biologically outstanding 'ecoregions' (biogeographic units defined according to a range of ecological parameters and known as the *Global 200*) which are representative of the world's biodiversity. A number of marine ecoregions have been identified and these will serve to focus future WWF investment in MPAs[126]. These various regional initiatives will help to ensure that MPA systems, at both national and regional levels, are designed on the best scientific information available, and that 'connectivity' is taken into account – that is the linkages that are created between areas through the dispersal and migration of larvae, juveniles, and adults of many marine species. The protection of 'sources' and 'sinks' of larvae, for example, may be particularly important[127].

Most existing MPAs are close, if not adjacent, to shore, but there is a growing need to protect areas of ocean and seabed, such as seamounts and hydrothermal vents, that lie offshore, as threats from shipping, fishing, and mineral exploitation escalate. Through IMO and the MARPOL Convention (see below – 4.4), open ocean areas can be designated for protection from pollution from ships, and through the IWC (see below – 2.3), sanctuaries to protect whales from exploitation can be set up. Areas of the seabed can be designated off-limit for mineral extraction through application of some of the provisions of UNCLOS. However, there is no international mechanism for establishing MPAs for general biodiversity protection on the high seas[128]. Some of the regional agreements include such provisions (such as the Antarctic Treaty Environmental Protocol and the Barcelona Convention) and others may follow, providing useful models for a more global system.

1.3 Improve the management of MPAs by:

- **preparing and implementing management plans**
- **building capacity for MPA management at local and national levels**
- **developing sustainable financing mechanisms**
- **establishing programmes for research, monitoring, and evaluation within MPAs.**

Improving the management of existing MPAs is as important as (and in some areas more important than) establishing new ones. Fewer than 50 per cent of existing MPAs are considered to be effectively managed[129]; too many are nothing more than 'paper parks', pin-up projects that look good on paper but fail in implementation. Reasons for this lack of effectiveness include inadequate enforcement of regulations, insufficient financial, technical, and human resources, lack of institutional support, and lack of public support due to failure to involve stakeholders. IUCN and WWF have been active in providing guidance and technical assistance, and IUCN has produced many of the key guidelines and handbooks[130].

All MPAs need proper and effective management plans that can be translated into implementable programmes, and in many cases these are lacking. As with protected areas on land, the management of MPAs can be contentious and complicated, and must take into account the needs and concerns of a diverse group of stakeholders. Involvement of local communities and major resource users at all stages is essential. In the Pacific, efforts are under way to strengthen traditional management practices, recognizing that these may provide one model for modern MPA management. With their unique knowledge of their own areas, local communities can play an important role in enforcement and monitoring.

For long-term effectiveness, MPAs must also be established within the context of an ICM plan which takes into account the fluid nature of the marine environment, the impacts that originate far away such as land-based activities and sources of pollution, and the migratory characteristics of species such as turtles, seabirds, and whales.

A secure financial basis is another prime requirement for the long-term success of an MPA, and a number of models and options for this are being investigated. Marine-based tourism is starting to pay for MPA upkeep in many parts of the world. The Great Barrier Reef Marine Park in Australia, for instance, attracted over 1.5 million visitors in 1994, and was generating some US$1.5 billion a year in the early 1990s[131]. Many MPAs, even small ones, can become self-sufficient

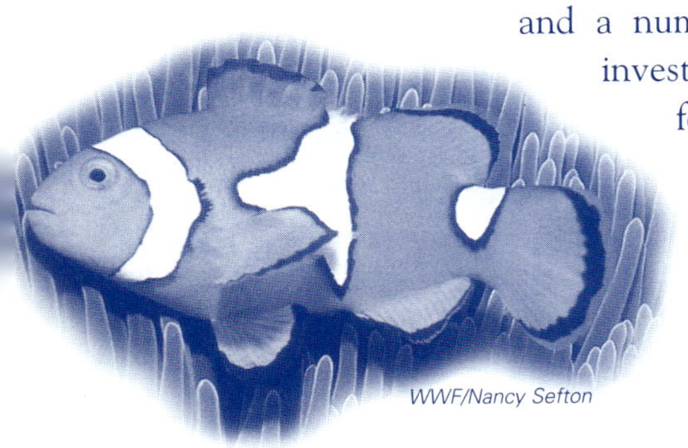

WWF/Nancy Sefton

entities, particularly if they have potential for well-managed diving and snorkelling (see box).

The establishment of research, monitoring, and evaluation programmes is also essential, if progress is to be measured, changes in the health of ecosystems identified, and management adapted to altered circumstances. At present, there is no globally accepted system of assessment and verification of management effectiveness in protected areas, either terrestrial or marine. This will require the development of an international set of standards that will take into account the need to include, in any national or regional system, a minimum area under strict protection designed to ensure the preservation of biodiversity, as well as a balanced range of multiple-use zones.

Bonaire's Marine Park

The coral reef fringed islands of Bonaire and Saba, in the Dutch Antilles in the Caribbean, illustrate how carefully managed marine protected areas can pay for themselves. With support from WWF, the waters surrounding both islands have been designated Marine Parks. Divers and snorkellers are required to pay US$10 a year in Bonaire and divers pay US$3 a dive in Saba. Permanent mooring buoys have been installed to eliminate anchor damage to the reef, and regular patrols ensure that park regulations are being observed.

The diving fees, combined with voluntary donations from support groups (largely made up of overseas visitors), now generate enough to cover the costs of managing the parks. In addition, tourists spend thousands of dollars on accommodation, food, support services, and souvenirs, providing income to the island communities. Tourists and tour operators alike approve of the fee system and the management plan, claiming that the islands' pristine reefs are now better managed[132].

OBJECTIVE 2: The conservation and recovery of threatened marine species.

Meeting the needs of threatened species is an important part of the ecosystem approach to marine conservation. In particular, so-called 'keystone' species such as large predators, which have a particular role in the maintenance of certain ecosystems, may need special attention. The high value of many marine species, for commercial, subsistence, and cultural reasons, also requires that particular attention be paid to them. In addition, many of the large marine species provide visible symbols of the value of the oceans and can play a special role in their conservation, engaging the public's attention and imagination. What the panda and the tiger accomplished for land-based conservation initiatives, the beluga, monk seal, bluefin tuna, giant clam, and others can do for marine efforts. The management challenge is twofold – to conserve and manage both species that are clearly connected to specific areas of sea or shore and those that are highly migratory.

Efforts to establish and manage MPAs, many of which encompass critical feeding and breeding habitats of threatened species, and activities to promote sustainable fisheries management (for example, by reducing bycatch of non-target species), make an important contribution and are covered under Objectives 1 and 3. This Objective draws attention to specific additional measures that are needed to ensure that all marine biodiversity is adequately protected. A global approach is necessary given the transboundary nature of both the ranges of many of the rarest species and the main threats that are affecting them, particularly international trade.

Priority activities

2.1 Assist in the identification of priorities for action to conserve threatened marine species by supporting the continuing development and application of:

- the IUCN *Red List* for marine species
- similar regional and national lists, through research, workshops, and other appropriate activities.

Work to identify threatened plant and animal species is undertaken by IUCN's Species Survival Commission (SSC), comprising 7,000 volunteer scientists, field researchers, natural resource managers, government officials, and conservation leaders from around the world who make up the Specialist Groups with responsibility for different taxonomic categories. This global network of experts helps to compile the IUCN *Red Lists* of threatened animals and plants. Representation of marine fish and invertebrates has so far been poor (see Disappearing species – page 28) but work to rectify this is being initiated and will involve looking closely at the criteria used for determining threatened species, and their application to marine species. Regional and national lists are paying increasing attention to marine species, examples being those for Peru and Brazil.

2.2 Support the preparation and implementation of Species Action and Recovery Plans.

Once threatened species have been identified, Action Plans must be drawn up specifying priority activities for their protection and recovery. The IUCN/SSC Specialist Groups for cetaceans[133] and seals[134] have produced such plans; others are under way (e.g. for molluscs) and the Specialist Groups for sharks, coral reef fish, and other marine fish are at the earlier stage of identifying the species to be considered. A global strategy has been produced for turtles[135]. Many of the activities under way through WWF, IUCN, and other organizations, both governmental and non-governmental, contribute to the implementation of these plans, but further, well coordinated efforts are needed to halt the decline of many marine species. The Action Plans provide an invaluable tool for ensuring that priority actions are undertaken; their successful implementation is illustrated by the fact that the Action Plan for cetaceans had to be revised only five years after its first edition, many of the recommended actions having been carried out. Recovery Plans for fish species are now required under certain regional fishery agreements (see below – 3.1).

2.3 Reduce exploitation of threatened species and demonstrate activities that involve their sustainable use by:

- monitoring and regulating international trade in marine species, using global mechanisms such as the Convention on International Trade in Endangered Species of Wild Fauna and Flora (CITES) and the IWC
- demonstrating and promoting sustainable use of marine species through non-consumptive activities, including whale watching, and traditional forms of sustainable subsistence use.

Exploitation of the economic and cultural value of many marine species has been the cause of their decline and extinction. Thus WWF and IUCN, while supporting the sustainable use of marine wildlife and the maintenance of traditional human lifestyles dependent on it, seek to eliminate forms of exploitation that threaten the long-term survival of any species.

The burgeoning international trade in many marine products is a particular threat and is poorly monitored and regulated at present. CITES is the main global mechanism for regulating international trade in threatened species, and the parties have endorsed the need for the convention to address unsustainable trade in commercial fishery species. However, although some of the *Red List* marine species are also listed in the Appendices (see table – page 29), the majority are terrestrial[136]. The only marine fishery species listed are invertebrates (giant clams and the queen conch). Proposals have been tabled to list bluefin tuna and certain sharks, but to date these have been rejected. TRAFFIC (Trade Records Analysis of Flora and Fauna in Commerce), the wildlife trade monitoring arm of WWF and IUCN, is undertaking a major fisheries trade programme, carrying out assessments of trade in fisheries products from species that are of particular concern, such as sharks[137], tuna[138], and seahorses[139]. This will help to identify those species which need more careful trade monitoring and regulation.

The conservation status of cetaceans is of particular concern, and there is an urgent need to minimize the full range of human impacts on this group, including pollution, habitat degradation, and bycatch in fisheries, as well as overexploitation. The legitimate take of whales for subsistence purposes by indigenous peoples with strong cultural ties to this practice should only be conducted on a strong scientific basis that ensures protection of endangered whale populations. The IWC, the body for worldwide regulation of whaling, has agreed to a moratorium on all commercial whaling. If this ban were to be lifted, it would be essential that any future commercial whaling be conducted on a strong scientific basis under international management and strict control by the IWC[140].

One way to take pressure off threatened species is through non-consumptive uses. The current growth in marine tourism, and the enthusiasm of many people to see marine species in the wild, particularly the large and charismatic ones, provides a promising opportunity in some areas. Whale watching, for example, has been shown to contribute higher economic returns than whaling (depending on the species and their status) (see box). Such activities, however, must be subject to appropriate controls, and there is a continuing need for guidelines. Giant clams provide an example of a group of threatened species that can

Whale watching

Tourist demand for whales, dolphins, and seals is proving almost insatiable. By 1994, whale watching was an attraction in 65 states and territories, there were 5.4 million whale watchers, and the industry was generating US$504 million annually[141]. Whale watching can change attitudes about whaling and provide an alternative source of income for whaling communities, and is becoming popular in some of the ex-whaling nations, as well as at many other tourist destinations throughout the world. In Iceland, for example, over 20,000 people paid to watch whales in 1997, earning the country some US$7 million and generating much more revenue than could be expected from exports of whale products. Husavik, where WWF is providing funding for a museum about whales and the history of whaling, is one of the main whale-watching towns, where tourists can observe up to seven species, including the blue whale.

The success in Iceland has led the national airline, the national tourist council, and the trade council all to come out publicly against the resumption of whaling. To ensure that this new 'industry' remains sustainable, however, guidelines and codes of conduct are essential to guard against adverse impacts on the whales themselves, and in some areas mandatory regulations may be needed.

benefit both from non-consumptive use (divers and snorkellers delight in discovering these impressive bivalves), and also from careful management (through ranching and establishing sanctuaries) for subsistence use by local communities in the Pacific who are traditionally dependent on these species[142].

2.4 Improve the understanding of, and work to mitigate, the impact on marine species of generic global threats such as:

- pollution from toxic chemicals
- commercial fisheries
- the introduction of alien species
- global climate change.

Many marine species have vast ranges compared with terrestrial species, or are migratory and spend key parts of their lives in widely different habitats. As a result, they are particularly at risk from factors that impact the entire marine ecosystem, such as toxic chemicals and global climate change. The conflict which can arise between threatened species, including whales and seals, and their food sources when these are of commercial value to humans is also a particular concern. Further work is required to understand the long-term effects of these impacts, but enough is already known – for example regarding cetaceans and toxics or the devastating effects of alien species – to suggest that these are serious. As understanding of the need to take a precautionary and ecosystem approach to fisheries and species management increases, it may become easier to find solutions to such problems.

OBJECTIVE 3: The introduction of measures to ensure that fishing is carried out in a sustainable manner, in order to conserve genetic, species, and ecosystem diversity.

The world needs nothing less than a 'blue revolution' to ensure the sustainable management of marine living resources. Sustainable fishing could bring many benefits, including a guaranteed source of high-quality protein for the world's poorest billion. Saving fisheries will also spill over into other areas, including the conservation of primary habitats. Conservation, restoration, and sustainability must therefore become the goals of modern fisheries management. *Ad hoc* measures such as gear restrictions (e.g. mesh size limits) and species-specific quotas (based on the outdated concept of 'maximum sustainable yield') have largely proved inadequate and ineffective. It is now recognized that managing fisheries has less to do with managing fish than with managing people who fish.

Governments must take the lead in developing and enacting economic incentives for sustainable fishing, thus providing a 'carrot' rather than merely the 'stick' of prescriptive regulations. In many cases, the first step will be to limit fishing effort so that depleted populations can recover to healthy levels. Market-driven incentives must also be created to harness market forces and consumer power behind efforts to recover and sustain well-managed fisheries and aquaculture.

WWF/Denis-Huot/BIOS

WWF and IUCN will identify mechanisms, best practices, and policy tools for both artisanal and commercial fisheries, and for aquaculture, which will promote the recovery of depleted populations, prevent damage to non-target species and habitat, encourage sustainable use of fisheries, and safeguard ecosystem integrity by:

- strengthening national, regional, and international capacities for fisheries management through training and institutional arrangements
- developing management approaches and regimes that encompass conservation objectives and socio-economic considerations, and that ensure equity in benefits
- increasing public awareness of the world fisheries crisis and disseminating information on potential solutions; this is crucial to building up political will and stimulating fundamental reforms
- promoting transparency, dialogue, and participation in fisheries management, planning, and decision making by developing policies and projects with all interested parties including fisheries scientists, managers, fishing communities, NGOs, businesses, and governments.

Priority activities

3.1 Promote conservation, species recovery, and the ecosystem approach as essential principles of fisheries management by:

- encouraging ratification and implementation by all countries, especially the world's top fishing nations, of the United Nations Agreement on Straddling Fish Stocks and Highly Migratory Fish Stocks (Fish Stocks Agreement)
- encouraging implementation of the FAO Code of Conduct for Responsible Fisheries
- facilitating the preparation and implementation of recovery plans for depleted species, especially those that serve as keystone species, such as tuna and sharks
- encouraging the appropriate establishment of NFZs within MPAs or as stand-alone measures.

The ecosystem approach to fisheries management is now being widely advocated[143]. This means taking into account all the parameters on which a fish stock depends, such as spawning, nursery, and feeding areas, as well as interactions with other species such as prey and predators, and adapting management measures if any aspect of the functioning of the ecosystem starts to fail.

Two mutually reinforcing initiatives, if fully implemented, could go a long way towards achieving this new approach. The Fish Stocks Agreement, adopted by the United Nations General Assembly in 1995, sets new international standards for the conservation and responsible management of all marine fisheries, and clarifies the provisions of UNCLOS in this respect, particularly in relation to species such as sharks, tuna, and billfish. By mid-April 1998, 59 countries had signed the agreement and 18 had ratified or acceded, although the treaty will not come into effect until this figure has reached 30.

The second key initiative is the Code of Conduct for Responsible Fisheries which was adopted at an FAO conference in November 1995. It is much broader in scope than the Fish Stocks Agreement, covering all fisheries throughout the world's seas, all aspects of fisheries management and fishing operations, aquaculture development, the integration of fisheries into coastal area management plans, post-catch practices and trade issues, and fisheries research. The Code of Conduct, however, is not legally binding.

Implementation and enforcement of the Fish Stocks Agreement and the Code of Conduct will be the responsibility of governments and the regional fisheries bodies. These will have to make fundamental changes in their operations if they are to be successful, for example allowing greater NGO participation and adopting a more biodiversity-oriented perspective on fisheries management. The Convention on the Conservation of Antarctic Marine Living Resources (CCAMLR), which provides a framework for improved fisheries management in the Southern Ocean, is one example in that it is the first international agreement to have applied an ecosystem approach to high-seas fisheries management.

For many species that have been subject to chronic overexploitation, the first step towards sustainable fishing will be the recovery, or rebuilding, of depleted populations to their former levels of abundance. Many domestic laws, such as the US Sustainable Fisheries Act of 1996, and regional fisheries agreements such as the 1994 Convention for the Conservation of Southern Bluefin Tuna (CCSBT) and the International Commission for the Conservation of Atlantic Tunas (ICCAT), mandate the development of recovery plans. These should incorporate the precautionary approach and require reduction in bycatch, identification and conservation of essential fish habitat, and re-evaluation and modification as appropriate of catch limits, quotas, minimum size limits, and other similar measures. Realizing the long-term benefits of species recovery plans usually requires politically unfavourable short-term sacrifices, which is undoubtedly why they have been difficult to prepare.

MPAs are an essential component of the ecosystem approach to fisheries management. They safeguard areas of high productivity that form the basis of fisheries and (in the case of multiple-use MPAs) can be used to help with fair allocation of fishery rights, including the historical rights of indigenous peoples. In addition, areas in which fishing is strictly limited or NFZs, either as stand-alone measures or as zones within multiple-use areas, have been shown to enhance fisheries yields in surrounding waters by protecting habitat critical to breeding populations as well as recruitment and spawning grounds. NFZs serve as an 'insurance policy' against overfishing, and give depleted fish

Fish sanctuaries in the Philippines

Many Philippine fishing communities are firm believers in the concept of 'no-fishing zones' (NFZs). Since the mid-1980s, a growing number of small marine protected areas (MPAs) have been set up by fishing villages, supported by the local governments, with the principal aim of better management of the coral reef fish populations on which they depend. The general pattern of these MPAs is of one or more 'sanctuary' areas where all fishing is banned, surrounded by a larger area in which local fishers only may fish using non-damaging methods, such as hook and line.

The MPAs are managed by the villages, often through the establishment of a committee. Fish populations and catches have been monitored within and around many of these areas, by the fishers themselves and by visiting scientists. At Sumilon, Pamilacan, Balicasag, and Apo Islands, densities of important food fishes such as snappers and groupers have been found to be 1.5 to 20 times as high within the sanctuaries as outside. These protected breeding populations ensure a rich supply of larvae to replenish the surrounding coral reefs that are fished. Increasing the argument for such NFZs, there is now evidence from Apo Island that as populations within the closed area build up, adults and juveniles emigrate out to surrounding reefs, increasing stocks there even further. Furthermore, fishers on the island are unanimous that their yields have increased since the reserve was established[144].

populations a chance to recover. Recent scientific work has shown that, once an area is protected, fish stocks build up rapidly and the size of target species increases (see box – page 43). The larvae of protected stocks are carried outside the boundaries of the NFZ to replenish fishing grounds, and adults and juveniles may migrate across the boundaries as stocks build up inside[145].

3.2 Reform economic and social policies which have led to overfishing by:

- reducing overcapacity through collaborative mechanisms developed with relevant partners, and by cutting harmful subsidies and encouraging the decommissioning of excess fishing fleet capacity
- developing mechanisms to reduce the impact of Northern fisheries on Southern countries, including codes of practice and stricter conditions for distant-water fishing agreements
- assisting in the development of valid principles, criteria, and indicators of sustainable fisheries
- supporting voluntary, independent certification and labelling initiatives based on such criteria
- supporting efforts to develop sustainable, equitable, and socially acceptable alternative livelihoods and economic options.

Perhaps the most important action that can be taken to improve the sustainability of fisheries is to reduce overcapacity and cut the massive subsidies to commercial fisheries[146]. These key issues are now being addressed by several of the international fora concerned with fisheries management, including FAO and the CSD. WWF and IUCN will contribute to these activities.

Quantitative reductions alone may not be appropriate. It will probably be important to identify fisheries-sector aid programmes that exacerbate fleet overcapacity, as well as financial assistance programmes that subsidize distant-water fishing fleets operating in the waters of developing nations and conflict with international law and policies for fish stock conservation. Such fishing often adversely affects small-scale and artisanal fishers who use the same resources. Ultimately, a code of practice will be needed for fishing by distant-water fleets. Subsidies reduction involves trade and social issues as well as fisheries and conservation concerns, and must therefore be addressed by a wide range of bodies[147]. This will require a mechanism to ensure collaboration between the CSD, the Committee on Trade and Environment of the WTO, UNEP, FAO, the CBD, and others.

WWF-Canon/Michel Gunther

Marine Stewardship Council (MSC)

The MSC is an independent, non-governmental, non-profit organization established in 1997 at the initiative of WWF and Unilever, a major seafood processor*. The MSC seeks to use market incentives and consumer power to encourage sustainable fishing, and already has the registered support of 43 fishing organizations, governments, fish retailers, academic institutions, and NGOs. Central to its effectiveness is a set of Principles and Criteria for Sustainable Fishing. An extensive international consultative process, which has to date included eight regional workshops, enables scientists, representatives of the fishing industry, NGOs, indigenous people, and other stakeholders to present their views. Draft principles and criteria are being tested in several fisheries, including rock lobsters in Western Australia, reef fishes in Eritrea, and herring in the Thames estuary (UK), prior to revision. The final Principles and Criteria will provide standards for a voluntary certification programme involving accredited certifiers independent of the MSC. These will evaluate fisheries for conformance with the MSC Principles and Criteria on behalf of a wide variety of clients, who will use certification for labelling fish products as 'sustainably harvested'.

*IUCN is not involved in the MSC.

The private sector also has a role to play in promoting sustainable fisheries. Increasingly, independent certification is seen as one of the most promising non-regulatory tools for helping to drive economic activities in the direction of sustainable development[148]. Certification and ecolabelling aim to link 'green consumers' to producers who are trying to develop sustainable management practices. However, before such processes can be initiated, a globally acceptable set of principles and criteria must be developed (see MSC panel – page 44). Certification and ecolabelling cannot replace other management tools, such as national fisheries policy, legislation, and education, but as more consumers opt to buy environmentally friendly products, certification can play an important complementary role and can help to shape policies, provided it can be implemented fairly in both Northern and Southern countries.

Artisanal fisheries programme in Guinea Bissau

In cooperation with local partners, IUCN is working in Rio Grande de Buba, an important but vulnerable estuary area in Guinea Bissau, where the artisanal fishery depends on local spawning and nursery areas, but has been threatened by overexploitation of stocks by large numbers of foreign fishers. Systems of credit and operating funds have been introduced, advice and assistance provided for fish processing and marketing, and for the development of vegetable gardens and crop cultivation to ensure both additional sources of livelihood and preservation of the forest in the catchment area. Stakeholder committees have been established, responsible for monitoring and enforcement, and training has been provided in numerical literacy, health, and sanitation.

The direct involvement of the local communities has ensured that new practices were implemented, and that legislation to recognize the exclusive access rights of the local fishers was subsequently adopted by the national government.

Regulatory controls, market-based incentives, and reductions in subsidies will have limited impact on problems at the small-scale or subsistence fishery level, and government institutions rarely have the capacity to address these complex, labour-intensive, low-capital activities. In such situations, the only successful approach seems to be that of community involvement and participation (see box above). The fundamental reform needed is that of empowering fishing communities, helping them develop a sense of ownership or partnership in the management process and, where relevant, reasserting previously held rights over fish stocks. As with MPA's, inspiration can be drawn from community-based traditional tenure systems of the Pacific, where defined fishing grounds and strict locally generated limitations helped to prevent overfishing. In other parts of the world, where such tenure systems do not exist, collaborative management between government, fishers, and NGOs may be a solution. In many cases, the establishment of a fishing cooperative has helped to engender the stewardship approach: although many cooperatives were initially set up to ensure equity in relation to income from catches, they are becoming increasingly involved in the maintenance of the quality of the fishery itself[149].

3.3 Reduce the negative impacts of fishing gears and practices on marine habitats and species by:

- strengthening enforcement of laws that ban destructive fishing practices, such as fishing with large-scale driftnets, poisons, and explosives
- promoting the use of bycatch reduction devices, including TEDs and other appropriate gear modifications
- developing alternative, environmentally friendly fishing methods, and training fishers to use them.

It is crucial that the negative effects of inappropriate and destructive fishing gears be addressed. In the Philippines, NGOs and citizens' action groups are working with cyanide and dynamite fishers in an effort to encourage them to adopt safer, more ecologically sound fishing methods. Showing how to capture live reef fish for the aquarium trade using nets, instead of poison, has had some notable success in this country and training efforts are now being extended to Indonesia. It is essential, however, that entire communities support the initiatives, as destructive practices will otherwise quickly replace more sustainable ones.

One of the most difficult and contentious issues is how to reduce the bycatch from commercial operations. Shrimp or prawn trawl fisheries account for nine of the ten fisheries with the world's worst bycatch records[150]. However, such bycatch is not an inevitable consequence of fishing. Shrimpers in the Gulf of Mexico are now required to have TEDs on their nets; when used properly, these greatly reduce the bycatch of marine turtles and other species. Similarly, a wide array of mechanisms that do not compromise fishing efforts has been developed to reduce seabird bycatch in long-line fisheries. Concerted efforts must be invested in educating fishers to use bycatch reduction techniques, developing new ones where needed, establishing incentives for their use, and expanding observer programmes to ensure their correct application.

If bycatch and waste cannot be reduced to acceptable levels, alternative more ecologically acceptable fishing methods must be found. In the early 1990s, the United Nations General Assembly imposed a global moratorium on the use of large-scale pelagic driftnets (longer than 2.5 km), due to concerns over their excessive bycatch. Further assessment of the environmental impacts of specific gear types will be an important step towards achieving the goals of sustainable fisheries and the protection of non-target species and habitats.

3.4 Develop environmentally, economically, and socially sustainable aquaculture that does not damage the marine and coastal environment by:

- participating in the development of globally applicable principles and criteria that are widely acceptable, through a multi-stakeholder process
- encouraging the development of, and demonstrating, methods, techniques, and policies that contribute to sustainable aquaculture
- establishing mechanisms to ensure that guidelines and best management practices are implemented, including regulatory frameworks at all political levels, trade and investment mechanisms, and independent third-party certification systems.

With the depletion of wild fish populations, economic pressure is mounting to expand production through aquaculture and other enhancement techniques. Recognizing that such approaches will be essential for human well-being, fishery enhancement and restoration measures must be designed so that they harm neither the coastal environment nor the people living there, are both ecologically and socially sustainable, and minimize negative impacts such as pollution, the introduction of alien species, and user conflicts.

Many fora have been convened to develop principles and guidelines for sustainable aquaculture, but these have tended to be dominated by one-interest parties, such as industry or government. There is still a need to bring these players together to address this issue through a multi-stakeholder process. If such globally accepted guidelines can be produced, there would then be potential for introducing an independent third-party certification system. As with wild-caught fisheries, consumers can play an important role in encouraging sustainable methods of aquaculture by seeking out products that are certified as having come from environmentally and socially sustainable operations.

Much could be achieved if governments were to implement fully the requirements of Article 9 of the FAO Code of Conduct for Responsible Fishing and follow the recommendations for aquaculture laid out in the Jakarta Mandate of the CBD. There are few aquaculture operations that could not be made more sustainable by introducing production methods based either on traditional practices or on technological innovation. These include strict controls on the release of wastes; careful siting of farms and prohibiting the use of mangroves; encouragement of the use of indigenous species; improvements in feeding practices; the development of closed systems so that wild populations are not exploited for broodstock; and consideration of social impacts. Producer countries should adopt best practices, including measures to ensure that the industry bears the full cost of aquaculture, and international financial institutions, such as the World Bank, other multilateral development banks, and bilateral aid agencies that promote and subsidize this industry should review the social and environmental effects of their support. Similarly, private investors, both domestic and international, must direct finance to more eco-friendly methods. Important tools for achieving this include comprehensive environmental assessment and multi-stakeholder consultation prior to investment[151].

OBJECTIVE 4: The reduction and elimination of marine pollution from land-based and marine sources.

Combating pollution from both land-based and marine sources is one of the most difficult and contentious of all the issues confronting the blue planet, and progress has been agonizingly slow. This is despite the fact that reducing and ultimately eliminating pollution is essential for conserving marine biodiversity and the many livelihoods that depend on it. The establishment of MPAs and development of sustainable fishery regimes is pointless if they continue to be undermined by pollution. For example, there has been enormous investment in the protection and management of the Waddensee, Europe's most important coastal wetland. Nevertheless, its population of harbour seals underwent a major decline in the early 1990s, which was linked to increasing levels of toxic chemicals in their tissues as a result of feeding on contaminated fish[152]. Although populations have now stabilized, this illustrates how water quality is central to any marine management effort. Pollution prevention is not a luxury but an absolute necessity that countries ignore at their peril.

Agenda 21, adopted at UNCED, advocates the adoption of a precautionary and anticipatory approach rather than a reactive one to tackling marine pollution. Accordingly, the first priority must be to minimize polluting discharges at the source, and to develop strategies that address pollution in vulnerable and sensitive areas.

Better understanding and recognition of the need to retain and recycle wastes, to draw up contingency plans to deal with accidents, to develop and use clean production

WWF-Canon/Martin Harvey

technologies, and to consider all factors that contribute to the production and discharge of pollutants are essential both on land and at sea. Point or discrete sources of pollution dumped directly into coastal waters by industries and municipalities are, relatively, easier to identify and control. In many cases, appropriate technology exists to prevent such pollution.

WWF/JJan van de Kam

Much more difficult to control are non-point or diffuse sources, for example from diverse agricultural activities. One of the biggest challenges is to reduce the tremendous loads of pollution that reach the oceans via river systems, as run-off, and through the atmosphere as rain. Governments need to plan and implement pollution reduction strategies as quickly as possible as part of watershed management programmes. The Netherlands, for instance, found that the country's entire agricultural base would have to be altered in order to reduce contamination of groundwater and coastal waters from nitrates and phosphates in fertilizers[153].

This type of pollution is often the result of activities that are far removed from the sea, taking place in sectors which have no obvious or direct link with the marine environment. Nevertheless, a wide variety of low-cost, best management practices has been developed which needs demonstrating, and incentives must be identified to encourage their adoption in coastal states worldwide. Furthermore, where pollution is a result of the deliberate introduction of noxious substances to the environment, as in the case of pesticide application, alternative environmentally friendly pest control systems should be further developed and used.

Given that marine pollution of all types does not recognize international frontiers, global mechanisms are another essential tool. Under UNCLOS, states have agreed to strict measures to prevent, reduce, and control pollution of the marine environment. As with fisheries, this general obligation is refined and implemented through a number of other global and regional instruments and mechanisms described below.

Priority activities

4.1 Increase awareness among decision makers and the general public of the threat to marine biodiversity and human activities dependent on the sea from toxic chemicals such as endocrine disrupting chemicals (EDCs), persistent organic pollutants (POPs), and pesticides.

With growing evidence that toxic chemicals may represent a major threat to marine biodiversity, as well as to humankind, there is an urgent need to draw this to public attention and ensure that the implications are fully understood. *Our Stolen Future*[154], published by WWF and available in 12 languages, has had a significant impact in this regard. WWF's US-based Global Toxics Campaign will ensure that the issue remains in the public eye, and WWF, with IUCN, will encourage, support, and advocate for the urgent changes that are needed. Increasing awareness of the alternatives that are available is also important. This can be achieved by, for example, disseminating information on alternatives to particularly well-known chemicals such as dichlorodiphenyltrichloroethane (DDT), as in the report prepared by WWF-Canada and WWF-US[155].

4.2 Ensure that the issue of toxic chemicals is addressed in appropriate international fora including the following:

- the Intergovernmental Forum on Chemical Safety (IFCS) and the proposed agreement on POPs
- the Ramsar Convention
- the Global Programme of Action for the Protection of the Marine Environment from Land-based Activities (GPA)
- the CSD
- the pesticide risk-reduction programme of the Organisation for Economic Cooperation and Development (OECD)
- agreements under the UNEP Regional Seas Programmes and other regional agreements
- the proposed new global agreement to eliminate the use of tributyltin (TBT) in anti-fouling paints.

Under the guidance of the IFCS, an international treaty is to be negotiated to phase out the use of 12 of the most dangerous POPs (predominantly pesticides)[156]. This has the endorsement of over 100 countries, and it is hoped that negotiations will be concluded by the year 2000. The issue is also being addressed through other global mechanisms. They include: the Ramsar Convention, which has recognized the necessity for pollution prevention measures in order to maintain the health of wetlands, many of which are coastal; the GPA, which has introduced initiatives to reduce land-based sources of pollution (see below – 4.5); and the CSD. The OECD's programme is specifically working on screening, and on testing guidelines for chemicals to identify whether they may be endocrine disruptors. Similar initiatives are being undertaken through the UNEP Regional Seas Programmes. Important advances in recognition of the need for phase-outs have been made in Europe through the OSPAR Convention and the North Sea Conference. WWF and IUCN have been involved with many of these initiatives.

WWF-Canon/Tony Rath

Large quantities of TBT and other organotins still enter the sea directly from the anti-fouling paints on ships' hulls, despite a number of national and regional bans on the use of this chemical. For example, there is a Europe-wide ban on the use of TBT on ships smaller than 25 metres, put in place after the collapse of oyster farms due to exposure to this chemical. Efforts are now under way through IMO's Marine Environment Protection Committee and other bodies, including international organizations such as WWF, to institute a global ban on organotins, including TBT.

4.3 Reduce and eliminate the production and use of toxic chemicals by:

- promoting the introduction and implementation of government restrictions on, or phase-outs of, selected toxic chemicals and the development of national pesticide reduction programmes

- reducing reliance on chemicals such as DDT by identifying and promoting alternative non-chemical approaches at the national and community levels.

Our lack of knowledge and inability to quantify accurately the risks that toxic chemicals pose demands an umbrella approach to preventing or reducing their production and release. This involves eliminating and reducing production, use, and stockpiles, and providing adequate alternatives. Banning individual compounds is a useful step but alone is not sufficient, and overall reliance on these substances must be reduced. This means developing cleaner technologies, using safer, less toxic chemicals, and preferably finding non-toxic alternatives. Integrated pest management (IPM), which makes use of a combination of natural predators, crop rotation, and more environmentally safe pesticides to ensure production, has already had very encouraging results in those agricultural systems where it has been introduced. Countries such as Sweden, the Netherlands, and Denmark have reduced pesticide use by over 50 per cent[157].

Good demonstrations of the range of alternatives that now exist are still needed. For example, there are many ways to control mosquitoes and other disease vectors that pose a serious threat to human health apart from non-selective spraying with DDT, a practice that is still carried out in many countries. Insect breeding sites (such as tanks of standing water) can be removed or covered, people can be encouraged to use mosquito nets, and biological or at least less toxic insecticides can be employed. Programmes to reduce the use of DDT, whilst still maintaining environmental and human health, are proving successful in several countries, including the Philippines, Mexico, and Tanzania[158]. Alternatives to organotins

WWF/M Lindhard

such as TBT are also becoming available, with some of the more progressive paint manufacturers actively working to produce new types of anti-fouling paint.

Much greater effort must also go into developing environmental reporting systems for tracking sources of pollutants affecting the marine environment, and advocating the public's 'right to know', a principle that was adopted at UNCED. Pollutant Release and Transfer Registers (PRTRs) are one effective mechanism for achieving this. These systems allow the tracking of chemicals in products as well as their use, transfer, and release from industrial facilities. Such registers, and other inventories such as the US Toxic Release Inventory, help to identify problems at an early stage, spur the development of cleaner production methods, and provide a measure of accountability[159]. The labelling of foods and products that may contain toxic substances is a similar tool for achieving policy change. These activities will require the close involvement and cooperation of industry.

WWF/B&C Lang

4.4 Strengthen global mechanisms and agreements for the monitoring and reduction of pollution from shipping and offshore industries by:

- lobbying for ratification and implementation of the International Convention for the Prevention of Pollution from Ships, 1973, and its 1978 Protocol, jointly referred to as MARPOL 73/78, and the development of improved environmental procedures under IMO

- increasing understanding of the impacts and environmental cost of oil spills and disseminating information on methods for prevention and mitigation

- promoting the development of internationally accepted environmental principles for mining, mineral exploitation, and decommissioning activities

- promoting the identification and protection of areas that are particularly at risk from shipping and offshore oil and mineral exploitation.

Both WWF and IUCN have Consultative Status at IMO, the main United Nations agency dealing with shipping-related issues, as well as offshore oil and gas development, and have been working to encourage greater environmental awareness in its activities, largely through participation in the activities of IMO's Marine Environmental Protection Committee.

Much of IMO's work focuses on the implementation of MARPOL, which is the main global agreement for regulating the disposal of wastes at sea by ships. Five Annexes cover various substances and cargo loads (oil, noxious liquids carried in bulk, packaged substances, sewage, and garbage and plastics)[160]. Many MARPOL obligations are still very poorly implemented, partly because their success depends on adequate waste-reception facilities at ports and these are often lacking. One approach will be to develop the PRTR concept (see above) within the shipping industry. Risk assessment and risk-reduction strategies to minimize the likelihood of spills at sea, and promotion of zero discharge and environmental best-practice policies, are also essential.

WWF and IUCN are also working with IMO on other shipping-related issues such as the proposed phase-out of the use of TBT in anti-fouling paints (see above), and a proposed new Annex to prevent the introduction of alien species by regulating the discharge of ballast water. This would cover mechanisms to reduce the amount of untreated ballast water entering ports, for example by deballasting and reballasting on the high seas, and would encourage efforts to find ways of treating ballast water to remove alien species. Some countries, such as the United States, Canada, and Australia have already adopted control measures, but a global agreement would help to ensure that such actions are taken on a worldwide basis.

There is still little overall regulation of offshore industries such as oil, gas, and mineral exploitation and there is an increasing need for the development of globally acceptable environmental principles and standards. These should incorporate the precautionary approach, clean production technologies, waste disposal issues, decommissioning of operations, and restoration of disturbed habitat. A new UNEP initiative to establish an information forum on offshore oil and gas, which aims to increase awareness of impacts and knowledge of best practices, is one first step and is being supported technically by WWF.

Particularly Sensitive Sea Area, Cuba

In September 1997, the International Maritime Organization (IMO) identified its second Particularly Sensitive Sea Area (PSSA) – the Sabana-Camaguey Archipelago, off the north coast of Cuba. WWF had been assisting the Cuban delegation for the previous two years to get this proposal through. The archipelago consists of some 990 cays and islands, many mangrove-covered, which stretch over an area of 465 km^2, and are bordered to the north by a barrier reef.

A variety of protective measures can be enforced within a PSSA, including routing measures for shipping, compulsory pilotage schemes, and traffic management systems. The only other PSSA, the Great Barrier Reef, was established in 1990.

With the widespread recognition that certain areas of ocean may need special protection, mechanisms to protect especially sensitive areas from the impacts of shipping and offshore mining need to be encouraged and implemented. MARPOL allows for the establishment of Special Areas where regulations are particularly stringent, for example zero discharges of oily wastes under Annex I. A number of Special Areas have been designated under the different MARPOL Annexes, such as the Wider Caribbean (Annex 5), Mediterranean and Antarctic (Annexes I and 5), and Baltic (Annexes 1, II and V). The 1991 IMO Guidelines for the Designation of Special Areas and Identification of Particularly Sensitive Sea Areas (PSSAs) offer further opportunities to protect specific areas from the impact of shipping (see box)[161]. Similar attention must be paid to identifying areas of the seabed that should be set aside and protected from any future oil or mineral exploitation.

4.5 Raise awareness that non-toxic pollutants (nutrients, sediment, sewage) can have a detrimental effect on the marine environment, and promote activities that reduce these by:

- assisting in the development and implementation of relevant global and regional mechanisms such as the GPA and the UNEP Regional Seas Programmes
- promoting pollution-reduction programmes, including social and economic incentives that encourage clean production.

The GPA, adopted at the Washington Conference in 1995, is designed to elaborate on pollution measures outlined in UNCLOS[162]. A large number of United Nations bodies and programmes (such as UNEP's Regional Seas Programmes) are involved in its implementation, with the participation of organizations such as IUCN and WWF. The GPA calls for the creation of a clearing-house mechanism to facilitate the transfer of technology, data, monitoring methods, and best management practices that will measure, minimize, and prevent marine pollution. Strategies must be developed to address marine pollution at source, and criteria identified to evaluate their effectiveness. The GPA thus provides guidelines to both national governments and regional bodies on tackling land-based sources of pollution and, although not legally binding, they could have an important impact. Preliminary steps towards implementation are being taken at the regional level, for example in West Africa and the Arctic.

As with toxic chemicals and marine sources of pollution, the technology and measures for reducing nutrient enrichment and sedimentation are largely known, but there is often resistance to their application or implementation. The forces behind pollution are often related to the fact that the environmental and social costs of pollution are not included in the price of the goods and services that cause it. If these costs could be assessed, the regulatory frameworks and economic instruments necessary for altering both prices and the behaviour of industry and society could be identified, ultimately leading to a reduction in pollution. Methods to identify such costs are now being developed[163]. Partnerships with companies in the private sector must also be developed to promote and recognize clean production, and incentives provided to encourage 'clean' agriculture and industry, and environmentally sound sewage and solid-waste disposal in municipalities, villages, and tourism developments.

OBJECTIVE 5: The promotion of ICM as an underlying principle in the sustainable management of marine and coastal ecosystems.

In most countries, the major urban, industrial, and tourist activities are concentrated on the coast, with the result that many incompatible activities are increasingly competing for limited space and resources. Furthermore, as emphasized throughout this policy, activities taking place throughout a watershed – often far inland – can cause immense damage to coastal waters. In addition, the threat of sea-level rise in many parts of the world has led to conflicts between the protection of natural ecosystems and the construction of sea defences. Typically, government responsibilities for coastal management are fragmented among many different agencies, with little cross-reference between those in charge of shipping, tourism, fisheries, public works, port development, and environmental protection, and little communication with those in charge of sectors such as agriculture and forestry. This fragmentation results in widespread unplanned development, rapid urbanization, and unregulated discharges of pollutants into the sea.

With the growing understanding that MPAs and measures aimed at protecting individual marine species will be ineffective on their own, the ICM approach is increasingly being adopted. Essentially, this means managing the coastal

WWF-Canon/Siegfried Woldhek

zone and its watershed as a single unit, emphasizing the ecosystem approach. As much emphasis must be put on, for example, land-use planning, pollution control, and environmentally sound agriculture and building practices, as on establishing MPAs and fisheries management. To achieve successful integration, all those responsible for these activities, including government agencies, business and industry, scientists, farmers, fishers, and local people, need to be able to share their concerns and jointly develop solutions. Thus, ICM should integrate government with the community, science with management, and sectoral with public interests in preparing and implementing actions that combine investment in development with environmental conservation, and which adapt as conditions evolve.

Priority activities

5.1 Assist in the development, strengthening, and implementation of regional and global frameworks for ICM including:

- the CBD, the GPA, and the CSD
- the UNEP Regional Seas Programmes.

The CBD and the GPA have been recognized as major tools for enhancing coastal resource management. Under the Jakarta Mandate of the CBD (see box – page 33), ICM (referred to as 'integrated coastal area management' (ICAM)), is identified as a priority for meeting the obligations of the convention in relation to marine biodiversity. The convention's Secretariat and its Subsidiary Body on Scientific, Technical and Technological Advice (SBSTTA) have been given the task of identifying and establishing processes for providing advice on ICM[164]. In 1999, the CSD will make oceans and seas a sectoral theme for its work, and will focus on tourism as an economic sector. This will provide an opportunity to address the growing problems associated with coastal tourism, issues that can be tackled effectively only through ICM. The UNEP Regional Seas Programmes and other regional agreements also provide opportunities to promote ICM. In the western Indian Ocean region, for example, countries are being encouraged to develop this approach through several regional initiatives, including conferences at ministerial level, technical workshops, and input into the 1985 Convention for the Protection, Management and Development of the Marine and Coastal Environment of the Eastern Africa Region (Nairobi Convention)[165]. WWF and IUCN are involved in many of these initiatives.

5.2 Assist in the establishment of ICM programmes and strengthen their implementation by:

- developing guidelines, principles, and implementing mechanisms
- preparing training materials and building capacity for ICM.

Numerous generic theoretical and technical guidelines on ICM have been produced, such as the *Noordjwik Guidelines*, prepared with World Bank support, those produced by FAO, IUCN, and OECD, and others produced by commercial publishers[166]. The World Bank also has guidelines which have to be followed in projects funded by loans from the bank[167].

Regional ICM guidelines include those produced by UNEP for the Caribbean[168] and the Mediterranean[169]. In most cases, such generic models must be adapted to local environmental, economic, and social conditions. Thus WWF-Malaysia is producing a series of ICM handbooks in English and Bahasa Malaysia[170]. In some cases, guidelines are needed for particular sectors or issues, such as tourism. For example, WWF, with tour operators, representatives from indigenous communities, researchers, managers, and other conservation organizations, has developed principles for sustainable tourism in the Arctic, with codes of conduct for both tour operators and tourists[171].

Increasing capacity for ICM at national and local levels is vitally important. A number of institutions now run training courses, and both WWF and IUCN consider capacity building a central part of their work.

5.3 Increase political support and awareness of the need for ICM by:

- supporting the development of demonstration or model ICM projects
- identifying and disseminating information on lessons learned.

Although 90 countries (about half the countries in the world with coastlines) have ICM projects or programmes under way, it is probable that barely one-quarter have actually reached the implementation stage. Nevertheless, numerous small projects and some larger national-level ones have begun, including the IUCN-led project in Tanzania (see box – page 56). Documentation of the evolution of the ICM approach under different situations, and of longer-term projects, is urgently needed. One example is found in Turkey, where WWF and the National Society for the Protection of Nature (DHKD) have since 1988 been working to protect the few remaining turtle nesting sites in the Mediterranean. Several beaches were declared protected areas but, with the growth of tourism as a vital economic activity in the area, it has become necessary to develop a comprehensive management plan for this stretch of coastline, one which would lead not only to long-term protection of the turtles but would also address economic and social issues affecting the local people[173]. Other documented long-term ICM programmes include those in Ecuador[174] and Belize[175].

WWF/Jan van de Kamm

Methods for monitoring and evaluating the progress and success of ICM under different conditions are urgently needed if experiences gained in this process are to be effectively transferred to others. Some 30 ICM projects are being assessed in the European Union with this aim in mind, and a number of similar initiatives are under way through agencies such as the United Nations Development Programme (UNDP). WWF has produced a handbook on monitoring and evaluation which aims to ensure that these are considered at the very beginning of a project. The book is in part based on experience gained during the development of the PROARCA/Costas project, a joint project with WWF-US, the Nature Conservancy, and the University of Rhode Island, to develop policy and strengthen capacity for coastal management in Central America[176].

Tanga Coastal Zone Conservation and Development Programme

In 1994, with assistance from Irish Aid, the government of Tanzania worked with IUCN to establish an integrated coastal management programme in Tanga Region, the northernmost coastal region of Tanzania. The aim was to improve the capacity of government and community institutions to help the local people use their near-shore fisheries, coral reefs, and mangroves in sustainable ways, including restoring degraded environments.

The programme focuses on four main priority areas: coral reefs and reef fisheries management; mangroves, coastal forests, and wildlife; awareness, education, and training; and community development and participation. A pilot village programme supplying support and advice to government officers and the communities of three fishing villages has had extremely positive results: mangrove cutting has stopped, there is a voluntary mangrove replanting and weeding programme, and destructive fishing practices, including dynamiting, have declined, largely through the enforcement efforts of the villagers themselves. The villagers are now much more aware of coastal conservation concerns and have the preliminary skills required to protect and use their resources sustainably. There is also now a trusting and collaborative relationship between the communities and government officers, who received training in facilitating villagers to make their own resource-use management decisions, and learnt from the villagers in the process[172].

5.4 Ensure that considerations of climate change are incorporated in ICM planning, including the need to restore natural coastal processes and habitats where possible.

Climate change and its potential effects in terms of sea-level rise must be taken into consideration in ICM planning. The cost of erecting man-made structures to protect coastal areas from flooding as sea levels rise is likely to be immense (protection of the Netherlands from a 50-centimetre rise has been put at US$3.5 trillion). Such defences often do more harm than good. Rising sea levels may in fact offer opportunities to restore and recreate natural coastal and intertidal habitats. Furthermore, this threat emphasizes the need to maintain natural coastal defences such as wetlands and coral reefs. In many tropical areas, healthy coral reefs that grow and keep pace with sea-level rise are one of the best natural defences and this must be emphasized in ICM planning[177].

WWF-Canon/Tony Rath

Policy summary

GOALS

- To maintain the biodiversity and ecological processes of marine and coastal ecosystems.
- To ensure that any use of marine resources is both sustainable and equitable.
- To restore marine and coastal ecosystems where their functioning has been impaired.

OBJECTIVE 1

The establishment and implementation of a comprehensive, global network of ecologically representative, well-managed marine protected areas (MPAs) designed to conserve areas of high biological importance and productivity.

Priority Activities

1.1 Assist in developing, strengthening, and implementing regional and global agreements and mechanisms for the establishment and management of MPAs, such as:

- the Convention on Biological Diversity (CBD), the United Nations Convention on the Law of the Sea (UNCLOS), the Convention on Wetlands of International Importance (Ramsar Convention), and the Commission on Sustainable Development (CSD)
- the Regional Seas Programmes of the United Nations Environment Programme (UNEP) and other relevant regional agreements.

1.2 Ensure that this global network is truly representative by:

- evaluating and identifying gaps, and creating new MPAs to address these, e.g. for cetaceans, pelagic species, and the seabed
- promoting the concept of MPAs in offshore, transboundary, and international waters and identifying priority sites and mechanisms by which they might be established and enforced.

1.3 Improve the management of MPAs by:

- preparing and implementing management plans
- building capacity for MPA management at local and national levels
- developing sustainable financing mechanisms
- establishing programmes for research, monitoring, and evaluation within MPAs.

OBJECTIVE 2

The conservation and recovery of threatened marine species.

Priority Activities

2.1 Assist in the identification of priorities for action to conserve threatened marine species by supporting the continuing development and application of:

- the IUCN *Red List* for marine species
- similar regional and national lists, through research, workshops, and other appropriate activities.

2.2 Support the preparation and implementation of Species Action and Recovery Plans.

2.3 Reduce exploitation of threatened species and demonstrate activities that involve their sustainable use by:

GUIDING PRINCIPLES

- **All measures and practices to conserve marine biodiversity and ecological processes must take human needs into account.**
- **The concept of stewardship must be fostered through education and by increasing awareness at all levels.**
- **Communities must be empowered to protect and manage their marine and coastal resources, and local and national management capacity must be increased.**
- **Social and economic incentives for conservation and sustainable use must be created.**
- **The interconnectedness of the oceans must be recognized through appropriate cooperative transboundary and international mechanisms.**
- **The precautionary approach must be applied and an ecosystem-based approach to management fostered.**
- **All solutions must be based on the best available science and knowledge, with encouragement given to the acquisition, dissemination, and exchange of information.**

- monitoring and regulating international trade in marine species, using global mechanisms such as the Convention on International Trade in Endangered Species of Wild Fauna and Flora (CITES) and the International Whaling Commission (IWC)
- demonstrating and promoting sustainable use of marine species through non-consumptive activities, including whale watching, and traditional forms of sustainable subsistence use.

2.4 Improve the understanding of, and work to mitigate, the impact on marine species of generic global threats such as:

- pollution from toxic chemicals
- commercial fisheries
- the introduction of alien species
- global climate change.

OBJECTIVE 3

The introduction of measures to ensure that fishing is carried out in a sustainable manner, in order to conserve genetic, species, and ecosystem diversity.

Priority Activities

3.1 Promote conservation, species recovery, and the ecosystem approach as essential principles of fisheries management by:

- encouraging ratification and implementation by all countries, especially the world's top fishing nations, of the United Nations Agreement on Straddling Fish Stocks and Highly Migratory Fish Stocks
- encouraging implementation of the Code of Conduct for Responsible Fisheries of the Food and Agriculture Organization (FAO)
- facilitating the preparation and implementation of recovery plans for depleted species, especially those that serve as keystone species, such as tuna and sharks
- encouraging the appropriate establishment of no-fishing zones within MPAs or as stand-alone measures.

3.2 Reform economic and social policies which have led to overfishing by:

- reducing overcapacity through collaborative mechanisms developed with relevant partners, and by cutting harmful subsidies and encouraging the decommissioning of excess fishing fleet capacity
- developing mechanisms to reduce the impact of Northern fisheries on Southern countries, including codes of practice and stricter conditions for distant-water fishing agreements
- assisting in the development of valid principles, criteria, and indicators of sustainable fisheries
- supporting voluntary, independent certification and labelling initiatives based on such criteria
- supporting efforts to develop sustainable, equitable, and socially acceptable alternative livelihoods and economic options.

3.3 Reduce the negative impacts of fishing gears and practices on marine habitats and species by:

- strengthening enforcement of laws that ban destructive fishing practices, such as fishing with large-scale driftnets, poisons, and explosives
- promoting the use of bycatch reduction devices, including turtle excluder devices (TEDs) and other appropriate gear modifications
- developing alternative, environmentally friendly fishing methods, and training fishers to use them.

3.4 Develop environmentally, economically, and socially sustainable aquaculture that does not damage the marine and coastal environment by:

- participating in the development of globally applicable principles and criteria that are widely acceptable, through a multi-stakeholder process
- encouraging the development of, and demonstrating, methods, techniques, and policies that contribute to sustainable aquaculture
- establishing mechanisms to ensure that guidelines and best management practices are implemented, including regulatory frameworks at all political levels, trade and investment mechanisms, and independent third-party certification systems.

OBJECTIVE 4

The reduction and elimination of marine pollution from land-based and marine sources.

Priority Activities

4.1 Increase awareness among decision makers and the general public of the threat to marine biodiversity and human activities dependent on the sea from toxic chemicals such as endocrine disrupting chemicals (EDCs), persistent organic pollutants (POPs), and pesticides.

4.2 Ensure that the issue of toxic chemicals is addressed in appropriate international fora including the following:

- the Intergovernmental Forum on Chemical Safety (IFCS) and the proposed agreement on POPs
- the Convention on Wetlands of International Importance (Ramsar)
- the Global Programme of Action for the Protection of the Marine Environment from Land-based Activities (GPA)
- the CSD
- the pesticide risk-reduction programme of the Organisation for Economic Cooperation and Development (OECD)
- agreements under the UNEP Regional Seas Programmes and other regional agreements
- the proposed new global agreement to eliminate the use of tributyltin (TBT) in anti-fouling paints.

4.3 Reduce and eliminate the production and use of toxic chemicals by:

- promoting the introduction and implementation of government restrictions on, or phase-outs of, selected toxic chemicals and the development of national pesticide reduction programmes
- reducing reliance on chemicals such as dichlorodiphenyltrichloroethane (DDT) by identifying and promoting alternative non-chemical approaches at the national and community levels.

4.4 Strengthen global mechanisms and agreements for the monitoring and reduction of pollution from shipping and offshore industries by:

- lobbying for ratification and implementation of the International Convention for the Prevention of Pollution from Ships, 1973, and its 1978 Protocol, jointly referred to as MARPOL 73/78, and the development of improved environmental procedures under the International Maritime Organization (IMO)
- increasing understanding of the impacts and environmental cost of oil spills and disseminating information on methods for prevention and mitigation
- promoting the development of internationally accepted environmental principles for mining, mineral exploitation, and decommissioning activities

- promoting the identification and protection of areas that are particularly at risk from shipping and offshore oil and mineral exploitation.

4.5 Raise awareness that non-toxic pollutants (nutrients, sediment, sewage) can have a detrimental effect on the marine environment, and promote activities that reduce these by:

- assisting in the development and implementation of relevant global and regional mechanisms such as the GPA and the UNEP Regional Seas Programmes
- promoting pollution-reduction programmes, including social and economic incentives that encourage clean production.

OBJECTIVE 5

The promotion of integrated coastal management (ICM) as an underlying principle in the sustainable management of marine and coastal ecosystems.

Priority Activities

5.1 Assist in the development, strengthening, and implementation of regional and global frameworks for ICM including:

- the CBD, the GPA, and the CSD
- the UNEP Regional Seas Programmes.

5.2 Assist in the establishment of ICM programmes and strengthen their implementation by:

- developing guidelines, principles, and implementing mechanisms
- preparing training materials and building capacity for ICM.

5.3 Increase political support and awareness of the need for ICM by:

- supporting the development of demonstration or model ICM projects
- identifying and disseminating information on lessons learned.

5.4 Ensure that considerations of climate change are incorporated in ICM planning, including the need to restore natural coastal processes and habitats where possible.

References

1. **Couper, A.** (ed) (1989) *The Times Atlas and Encyclopaedia of the Sea.* Times Books, London, UK.

2. **Groombridge, B. and Jenkins, M.D.** (eds) (1996) *The Diversity of the Seas: A Regional Approach.* World Conservation Monitoring Centre, World Conservation Press, Cambridge, UK. 132pp. **Norse, E.A.** (1993) *Global Marine Biological Diversity.* Island Press, Washington D.C., USA. 383pp.

3. **Reaka-Kudla, M.L., Wilson, D.E. and Wilson, E.O.** (eds) (1996) *Biodiversity II: Understanding and Protecting our Natural Resources.* National Academy/ Joseph Henry Press, Washington D.C., USA.

4. **Grassle, J.F.** (1991) Deep-sea benthic biodiversity. *Bioscience* 41:464-469.

5. **WCMC** (1992) *Global Biodiversity: Status of the Earth's Living Resources.* Chapman and Hall, London, UK. 594pp.

6. **WWF** (1998) *Whales in the Wild.* WWF International, Gland, Switzerland.

7. **Lean, G. and Hinrichsen, D.** (1994) *Atlas of the Environment.* HarperPerennial, New York, USA.

8. **Thorne-Miller, B. and Catena, J.** (1991) *The Living Ocean – Understanding and Protecting Marine Biodiversity.* Island Press, Washington D.C., USA. 180pp.

9. **Horton, T. and Eichbaum, W.M.** (1991) *Turning the Tide: Saving Chesapeake Bay.* Island Press, Washington D.C., USA. 327pp.

10. **Spalding, M., Blasco, F. and Field, C.** (eds) (1996) *World Mangrove Atlas.* International Society for Mangrove Ecosystems, World Conservation Monitoring Centre, International Tropical Timber Organization, London, UK. 178pp.

11. **Fortes, M.** (1990) *Seagrasses: A Resource Unknown in the ASEAN Region.* ICLARM-USAID, Manila, Phillipines.

12. **Spalding, M.D. and Grenfell, A.M.** (1997) New estimates of global and regional coral reef areas. *Coral Reefs* 16:225-230.

13. **Craik, W.** (1992) The Great Barrier Reef Marine Park: its establishment, development and current status. *Marine Pollution Bulletin* 25:122-133.

14. **Lundin, C.G. and Linden, O.** (1993) Coastal ecosystems: attempts to manage a threatened resource. *Ambio* 22(7):468-473.

15. **Norse, E.A.** (1993) *Global Marine Biological Diversity.* Island Press, Washington D.C., USA. 383pp.

16. **Thorne-Miller, B. and Catena, J.** (1991) *The Living Ocean – Understanding and Protecting Marine Biodiversity.* Island Press, Washington D.C., USA. 180pp.

17. **Elder, D. and Pernetta, J.** (eds) (1991) *Oceans: A Mitchell Beazley World Conservation Atlas.* Mitchell Beazley, London, UK. **Kelleher, G., Bleakley, C. and Wells, S.** (1995) *A Global Representative System of Marine Protected Areas.* Vols 1-4. World Bank/ IUCN/Great Barrier Reef Marine Park Authority, Queensland, Australia.

18. **Weber, M. and Gradwohl, J.** (1995) *The Wealth of Oceans.* Norton & Company, New York, USA.

19. **Costanza, R. et al.** (12 other authors) (1997) The value of the world's ecosystem services and natural capital. *Nature* 387:253-260.

20. **McGinn, A.P.** (1998) *Rocking the Boat: Conserving Fisheries and Protecting Jobs.* Worldwatch Paper No. 142, Worldwatch Institute, Washington D.C., USA.

21. **Weber, P.** (1993) *Abandoned Seas: Reversing the Decline of the Oceans.* Worldwatch Paper No. 116, Worldwatch Institute, Washington D.C., USA. **Weber, P.** (1994) *Net Loss: Fish, Jobs, and the Marine Environment.* Worldwatch Paper No. 120, Worldwatch Institute, Washington D.C., USA.

22. **Loh, J., Rander, J., MacGillivray, A., Kapos, V., Jenkins, M., Groombridge, B., Cox, N.** (1998) *Living Planet Report.* WWF International, Gland, Switzerland. 36pp.

23. **Garcia, S. and Newton, C.** (in press) *Global Trends in Fisheries Management.* American Fisheries Society Monograph Series.

24. **Lundin, C.G. and Linden, O.** (1993) Coastal ecosystems: attempts to manage a threatened resource. *Ambio* 22(7):468-473. **Loh, J., Rander, J., MacGillivray, A., Kapos, V., Jenkins, M., Groombridge, B., Cox, N.** (1998) *Living Planet Report.* WWF International, Gland, Switzerland. 36pp.

25. **McGinn, A.P.** (1998) Promoting Sustainable Fisheries. Chap. 4. In: *State of the World 1998: a Worldwatch Institute Report on Progress Toward a Sustainable Society.* USA. pp 59-78. Norton, New York,

26. **McGinn, A.P.** (1998) Promoting Sustainable Fisheries. Chap. 4. In: *State of the World 1998: a Worldwatch Institute Report on Progress Toward a Sustainable Society.* pp 59-78. Norton, New York, USA.

27. **Elder, D. and Pernetta, J.** (eds) (1991) *Oceans: A Mitchell Beazley World Conservation Atlas.* Mitchell Beazley, London, UK.

28. **Porter, G.** (1988) *Resources, Population and the Philippines' Future: A Case Study.* World Resources Institute, Washington D.C., USA.

29. **Dugan, P.** (ed) (1996) *Wetlands in Danger: A Mitchell Beazley World Conservation Atlas.* Mitchell Beazley, London, UK.

30. **McGinn, A.P.** (1998) Promoting Sustainable Fisheries. Chap. 4. In: *State of the World 1998: a Worldwatch Institute Report on Progress Toward a Sustainable Society.* pp 59-78. Norton, New York, USA.

31. **IFRPI** (1997) Overfished Oceans, Booming Fisheries: What Does this Mean for World Food Security? *News & Views: A 2020 Vision for Food, Agriculture, and the Environment.* October 1997. International Food Policy Research Institute. Washington, D.C., USA. **McGinn A.P.** (1998) Promoting Sustainable Fisheries. Chap. 4. In: *State of the World 1998: a Worldwatch Institute Report on Progress Towards a Sustainable Society.* USA. pp 59-78. Norton, New York.

32. **FAO** (1997) *The State of World Fisheries and Aquaculture 1996.* FAO, Rome, Italy.

33. **Ratafia, M.** (1995) Aquaculture today: a worldwide status report. *World Aquaculture.* June 1995. **McGinn, A.P.** (1998) Blue revolution – the promises and pitfalls of fish farming. *World Watch* 11(2):10-19.

34. **Gujja, B. and Finger-Stich, A.** (1996) What price prawn? Shrimp aquaculture's impact in Asia. *Environment* 28(7). **FAO** (1997) *The State of World Fisheries and Aquaculture 1996.* FAO, Rome, Italy.

35. **Barber, C.V. and Pratt V.R.** (1997) *Sullied Seas: Strategies for Combating Cyanide Fishing in Southeast Asia and Beyond.* World Resources Institute/International Marinelife Alliance, Washington D.C., USA.

36. **Bentley, N.** (1998) An overview of the exploitation, trade and management of corals in Indonesia. *TRAFFIC Bulletin* 17(2):67-78; **Wood, E. and Wells, S.M.** (1995). The Shell Trade: A Case for Sustainable Utilization. In: Kay, E.A. (ed). *The Conservation on Biology of Molluscs.* pp 41-52. IUCN/SSC, Gland, Switzerland.

37. **Vincent, A.C.J.** (1996) *The International Trade in Seahorses.* TRAFFIC International, Cambridge, UK. 163pp.

38. **Wells, S. and Hanna, N.** (1992) *The Greenpeace Book of Coral Reefs.* Blandford, UK. 160pp.

39. **Fontaubert, C. de, Downes, D. and Agardy, T.** (1996) *Biodiversity in the Seas – Implementing the Convention on Biological Diversity in Marine and Coastal Habitats.* IUCN, WWF, CIEL, Gland, Switzerland. 82pp. **Fenical, W.** (1996) Marine biodiversity and the medicine cabinet: the status of new drugs from marine organisms. *Oceanography* 9(1):23-24.

40. **Groombridge, B. and Jenkins, M.D.** (eds) (1996) *The Diversity of the Seas: A Regional Approach.* World Conservation Monitoring Centre, World Conservation Press, Cambridge, UK. 132pp. **Norse, E.A.** (1993) *Global Marine Biological Diversity.* Island Press, Washington D.C., USA. 383pp.

41. **Field, C.D.** (1995) *Journey amongst Mangroves.* International Society for Mangrove Ecosystems. Okinawa, Japan. 140pp.

42. **Elder, D. and Pernetta, J.** (eds) (1991) *Oceans: A Mitchell Beazley World Conservation Atlas.* Mitchell Beazley, London, UK.

43. **Elder, D. and Pernetta, J.** (eds) (1991) *Oceans: A Mitchell Beazley World Conservation Atlas.* Mitchell Beazley, London, UK.

44. *The Economist* (1998) Special Survey: Travel and tourism, home and away. *The Economist* pp 1-16.

45. **Wheat, S.** (1997) The tourism juggernaut. *People and the Planet* 6(4):6-8.

46. **Batisse, M.** (1994) Mending the Med. *People and the Planet* 3(1):17-18.

47. **McField, M., Wells, S. and Gibson, J.** (eds) (1996) *State of the Coastal Zone Report, Belize 1995.* Coastal Zone Management Programme, Government of Belize.

48. **Batisse, M.** (1994) Mending the Med. *People and the Planet* 3(1):17-18.

49. **Hinrichsen, D.** (1998) *Coastal Waters of the World: Trends, Threats and Strategies.* Island Press, Washington D.C., USA.

50. **Vitousek, P.M., Mooney, H.A., Lubchenco, J. and Melillo, J.M.** (1997) Human domination of Earth's ecosystems. *Science* 277:494-499. **Hinrichsen, D.** (1998) *Coastal Waters of the World: Trends, Threats and Strategies.* Island Press, Washington D.C., USA.

51. **Hails, A.J.** (ed) (1996) *Wetlands, Biodiversity and Ramsar Convention.* Ramsar Convention Bureau/ Ministry of Environment and Forests, India. 196pp. **Spalding, M., Blasco, F. and Field, C.** (eds) (1996) *World Mangrove Atlas.* International Society for Mangrove Ecosystems, World Conservation Monitoring Centre, International Tropical Timber Organization, London, UK. 178pp.

52. **Horton, T. and Eichbaum, W.M.** (1991) *Turning the Tide: Saving Chesapeake Bay.* Island Press, Washington D.C., USA. 327pp.

53. **Eakin, C.M., McManus, J.W., Spalding, M.D. and Jameson, S.C.** (1997) Coral reef status around the world: where are we and where do we go from here? *Proceedings of the 8th International Coral Reef Symposium* 1:277-282. **Bryant, D., Burke, L., McManus, J. and Spalding, M.** (1998) *Reefs at Risk: A Map-based Indicator of Threats to the World's Coral Reefs.* World Resources Institute, Washington D.C., USA. 56pp.

54. **Agardy, T.** (1997) *Marine Protected Areas and Ocean Conservation.* Academic Press, Austin, USA. 244pp.

55. **FAO** (1995) *Review of the State of the World Fishery Resources: Marine Fisheries.* FAO Fisheries Circular No. 884. Rome, Italy. 105pp.

56. **Weber, P.** (1994) *Net Loss: Fish, Jobs, and the Marine Environment.* Worldwatch Paper No. 120, Worldwatch Institute, Washington D.C., USA.

57. **Dayton, P.K.** (1995) *Aquatic Conservation: Marine and Freshwater Systems.* Wiley, New York, USA.

58. **Rose, D.A.** (1996) *An Overview of World Trade in Sharks and other Cartilaginous Fishes.* A 'Species in Danger' report, TRAFFIC International, Cambridge, UK.

59. **WWF** (1996) *Marine Fishes in the Wild.* WWF Species Status Report, WWF, Gland, Switzerland.

60. **World Bank, UNDP, UNEP** (1993) *Saving the Black Sea, Programme for the Environmental Management and Protection of the Black Sea.* World Bank, UNDP, UNEP, Washington D.C., USA. pp 2-28. **McGinn A.P.** (1998) Promoting Sustainable Fisheries. Chap. 4. In: *State of the World 1998: a Worldwatch Institute Report on Progress Towards a Sustainable Society.* USA. pp 59-78. Norton, New York.

61. **Pauly, D., Christensen, V., Dalsgaard, J., Froese, R. and Torres, F.** (1998) Fishing down marine foodwebs. *Science* 279:860-863.

62. **Garcia, S. and Newton, C.** (in press) *Global Trends in Fisheries Management.* American Fisheries Society Monograph Series. **Porter, G. (1998)** *Too Much Fishing, Too Few Fish: A Proposal for Eliminating Global Fishing Capacity.*WWF-US/IUCN-US, Washington D.C., USA.

63. **Myers, N.** (1998) *Perverse Subsidies: Tax Dollars Undercutting Our Economies and Environments Alike.* International Institute for Sustainable Development. **WWF** (1997) *Subsidies and Depletion of World Fisheries: Case Studies.* WWF's Endangered Seas Campaign. WWF, Godalming, UK. 136pp.

64. **McGinn, A.P.** (1998) Promoting Sustainable Fisheries. Chap. 4. In: *State of the World 1998: a Worldwatch Institute Report on Progress Toward a Sustainable Society.* pp 59-78. Norton, New York, USA. **Weber, P.** (1994) *Net Loss: Fish, Jobs, and the Marine Environment.* Worldwatch Paper No. 120, Worldwatch Institute, Washington D.C., USA. **WWF** (1996) *Marine Fishes in the Wild.* WWF Species Status Report, WWF, Gland, Switzerland.

65. **Panos Institute** (1995) *Fish: a Net Loss for the Poor – Developing Countries and the Impact of Dwindling World Fish Stocks.* Panos Media Briefing 15, Panos Institute, London, UK. **WWF** (1996) *Marine Fishes in the Wild.* WWF Species Status Report, WWF, Gland, Switzerland.

66. **FAO** (1997) *Fisheries By-catch and Discards.* Fisheries Information Paper No. 7 to FAO Committee on Fisheries. FAO, Rome, Italy. **McGinn, A.P.** (1998) Promoting Sustainable Fisheries. Chap. 4. In: *State of the World 1998: a Worldwatch Institute Report on Progress Toward a Sustainable Society.* Norton, New York, USA. pp 59-78. **Alverson, D.L., Freeberg, M.H., Murawski, S.A. and Pope, J.G.** (1994) *A Global Assessment of Fisheries By-catch and Discards.* FAO Fisheries Technical Paper 339. FAO, Rome, Italy. 233pp.

67. **Alverson, D.L., Freeberg, M., Murawski, S.A. and Pope, J.G.** (1994) *A Global Assessment of Fisheries By-catch and Discards.* FAO Fisheries Technical Paper 339. FAO, Rome, Italy.

68. **Williams, T.** (1995) The Turtle Gulf War. *Audubon* 97(5):26-33.

69. **Alexander, K., Robertson, G. and Gales, R.** (1997) *The Incidental Mortality of Albatrosses in Longline Fisheries.* Australian Antarctic Division, Tasmania, Australia.

70. **Barber, C.V. and Pratt, V.R.** (1997) *Sullied Seas: Strategies for Combating Cyanide Fishing in Southeast Asia and Beyond.* World Resources Institute/International Marinelife Alliance, Washington, D.C., USA. **Johannes, R. and Riepen, M.** (1995) *Environmental, Economic, and Social Implications of the Live Reef Fish Trade in Asia and the Western Pacific.* The Nature Conservancy. **WWF** (1996) *Marine Fishes in the Wild.* WWF Species Status Report, WWF, Gland, Switzerland.

71. **Wells, S. and Hanna, N.** (1992) *The Greenpeace Book of Coral Reefs.* Blandford, UK. 160pp.

72. **McGinn, A.P.** (1998) Blue revolution – the promises and pitfalls of fish farming. *World Watch* 11(2):10-19. **Safina, C.** (1998) Scorched-earth fishing. *Issues in Science and Technology* 14(3):32-35.

73. **Clay, J.W.** (1998) Shrimp aquaculture: where are we headed? *Intercoast* 30:16-17.

74. **McGinn, A.P.** (1998) Blue revolution – the promises and pitfalls of fish farming. *World Watch* 11(2):10-19.

75. **McGinn, A.P.** (1998) Blue revolution – the promises and pitfalls of fish farming. *World Watch* 11(2):10-19.

76. **Folke, C. and Jansson, A.M.** (1992) *The Emergence of an Ecological Economics Paradigm: Examples from Fisheries and Aquaculture.* Beijer Reprint Series No. 9, Beijer International Institute of Ecological Economics, Royal Swedish Academy of Sciences, Kluwer Academic Publishers, Stockholm, Sweden.

77. **McGinn, A.P.** (1998) Blue revolution – the promises and pitfalls of fish farming. *World Watch* 11(2):10-19.

78. **McGinn, A.P.** (1998) Blue revolution – the promises and pitfalls of fish farming. *World Watch* 11(2):10-19.

79. **Pullen, S. and Hurst, P.** (1993) *Marine Pollution Prevention.* WWF International Background Report. WWF International, Gland, Switzerland. 34pp.

80. **FAO** (1990) *Water for Life.* FAO, Rome, Italy.

81. **Pearce, F.** (1995) Dead in the water. *New Scientist* 145(1963):26-31.

82. **UNEP** (1992) *Working Plan of North-West Pacific Region – China National Report.* United Nations Environment Programme, Beijing, China.

83. **Pullen, S. and Hurst, P.** (1993) *Marine Pollution Prevention.* WWF International Background Report. WWF International, Gland, Switzerland. 34pp. **WWF** (1997) Hormone disrupting chemicals in the environment – the threat to wildlife and humans. *Marine Update* 31, WWF-UK, Godalming, UK.

84. **Hinrichsen, D.** (1998) *Coastal Waters of the World: Trends, Threats and Strategies.* Island Press, Washington D.C., USA. **Pullen, S. and Hurst, P.** (1993) *Marine Pollution Prevention.* WWF International Background Report. WWF International, Gland, Switzerland. 34pp.

85. **Colborn, T., Dumanoski, D. and Peterson Myers, J.** (1996) *Our Stolen Future.* Abacus, UK.

86. **WWF** (1997) Hormone disrupting chemicals in the environment – the threat to wildlife and humans. *Marine Update* 31, WWF-UK, Godalming, UK.

87. **WWF** (1996) *Protecting Cetaceans from Hidden Threats: The Toxic Burden.* WWF, Gland, Switzerland.

88. **Pullen, S. and Hurst, P.** (1993) *Marine Pollution Prevention.* WWF International Background Report. WWF International, Gland, Switzerland. 34pp.

89. **WWF** (1997) Hormone disrupting chemicals in the environment – the threat to wildlife and humans. *Marine Update* 31, WWF-UK, Godalming, UK.

90. **FAO** (1995) *Effects of Riverine Inputs on Coastal Ecosystems and Fisheries Resources.* FAO Fisheries Technical Paper 349. FAO, Rome, Italy. 133pp. **Wasson, R.J.** (1997) Run-off from the Land to the Rivers and the Sea. In: *The Great Barrier Reef: Science, Use and Management. Proceedings of a National Conference* vol. 1 pp 23-41. Great Barrier Reef Marine Park Authority, Queensland, Australia.

91. **Helsinki Commission** (1994) *20 Years of International Cooperation for the Baltic Marine Environment 1974-1994.* Helsinki Commission, Helsinki, Finland.

92. **Hinrichsen, D.** (1998) *Coastal Waters of the World: Trends, Threats and Strategies.* Island Press, Washington D.C., USA.

93. **MacGarvin, M.** (1990) *The North Sea.* Collins and Brown, London, UK.

94. **Pullen, S. and Hurst, P.** (1993) *Marine Pollution Prevention.* WWF International Background Report. WWF International, Gland, Switzerland. 34pp.

95. **Steiner, R.** (1995) Probing an oil-stained legacy. *National Wildlife* 31(3):4-11.

96. **MacGarvin, M.** (1990) *The North Sea.* Collins and Brown, London, UK. **Hudgins, C.M.** (1991) *Chemical Useage in North Sea Oil and Gas Production and Exploration Operations.* Report for the Norwegian Oil Industry Association, Petrotech Consultants Inc., Houston, Texas, USA.

97. **WRI** (1995) *Coastlines at Risk: An Index of Potential Development-Related Threats to Coastal Ecosystems.* World Resources Institute, Washington D.C., USA.

98. **Pullen, S. and Hurst, P.** (1993) *Marine Pollution Prevention.* WWF International Background Report. WWF International, Gland, Switzerland. 34pp.

99. **Carlton, J.** (1998) Bioinvaders in the sea: reducing the flow of ballast water. *World Conservation* l:9-10. **Cohen, A.N. and Carlton, J.T.** (1998) Accelerating invasion rate in a highly invaded estuary. *Science* 279:555-557. **Boudouresque, C.F.** (1998) The great escape. *World Conservation* 1:17.

100. **IPCC** (1995) *Second Assessment Report.* Intergovernmental Panel on Climate Change.

101. **WWF** (1996) *Explaining Climate Change.* WWF, London, UK.

102. **Polovina, J.J., Mitchum, G.T., Graham, N.E., Craig, M.P., Demartini, E.E. and Flint, E.N.** (1994) Physical and biological consequences of a climate event in the central North Pacific. *Fish. Oceanogr.* 3(1):15-21. **Roemmich, D. and MacGowan, J.** (1995) Climatic warming and the decline of zooplankton in the California Current. *Science* 267:1324-1326. **Veit, R. et al.** (1997) Apex marine predator declines ninety percent in association with changing oceanic climate. *Global Climate Biology* 3.

103. **IPCC** (1995) *Second Assessment Report.* Intergovernmental Panel on Climate Change. **WWF** (1996) *Explaining Climate Change.* WWF, London, UK.

104. **Brown, B.E.** (1997) Coral bleaching: causes and consequences. *Proceedings of the 8th International Coral Reef Symposium* 1:65-74. **Bryant, D., Burke, L., McManus, J. and Spalding, M.** (1998) *Reefs at Risk: a Map-based Indicator of Threats to the World's Reefs.* World Resources Institute, Washington D.C., USA. 56pp.

105. **Agardy, T.** (1996) *Prospective climate change impacts on cetaceans and its implications for the conservation of whales and dolphins.* Paper presented at the IWC Symposium on Climate Change and Cetaceans, Hawaii, 1996. **Pernetta, J., Leemans, R., Elder, D., and Humphrey, S.** (1994) *Impacts of Climate Change on Ecosystems and Species: Marine and Coastal Ecosystems.* IUCN, Cambridge, UK, and Gland, Switzerland.

106. **IUCN** (1996) *1996 IUCN Red List of Threatened Animals.* IUCN, Gland, Switzerland. **Reaka-Kudla, M.L., Wilson, D.E. and Wilson, E.O.** (eds) (1996) *Biodiversity II: Understanding and Protecting our Natural Resources.* Joseph Henry/National Academy Press, Washington D.C., USA.

107. **WWF** (1998) *Whales in the Wild.* WWF International, Gland, Switzerland.

108. **Hayes, E.A.** (1997) *A Review of the Southern Bluefin Tuna Fishery: Implications for Ecologically Sustainable Management.* TRAFFIC Oceania and WWF, Sydney, Australia. 34pp.

109. **Earle, M.** (1996) Ecological Interactions between Cetaceans and Fisheries. Chap 7. In: Simmonds, M.P. and Hutchinson, J.D. (eds) pp 167-204. *The Conservation of Whales and Dolphins.* John Wiley and Sons Ltd, New York, USA.

110. **WWF** (1996) *Marine Fishes in the Wild,* WWF Species Status Report, WWF, Gland, Switzerland.

111. **Roberts, M.** (1998) *Lophelia pertusa* – a cold water coral. WWF UK Marine Factsheet. **Reaka-Kudla, M.L.** (1997) Biodiversity and its loss on global coral reefs: now and the next half century. *Tropical Diversity: Origins, Maintenance and Conservation.* Assoc. Tropical Biol. and Org. Tropical Stud, Abstract. **Malakoff, D.** (1997) Extinction on the high seas. *Science* 277:486-488. **Carlton, J.T.** (1993) Neoextinctions of marine invertebrates. *Amer. Zool.* 33:499-509.

112. **Hardin, G.** (1968) The tragedy of the commons. *Science* 162:1243-1248.

113. **SeaWeb** (1997) Majority of Americans say that Oceans should be priority over Space Exploration. Press release and graphics, SeaWeb – a project of the Pew Charitable Trusts.

114. **Christie, P. and White, A.T.** (1997) Trends in development of coastal area management in tropical countries: from central to community organisation. *Coastal Management* 25:155-181. **Borrini-Feyerabad, G.** (1996) *Collaborative Management of Protected Areas: Tailoring the Approach to the Context.* Issues in Social Policy, IUCN, Gland, Switzerland. **Jorge, M.A.** (1997) Developing capacity for coastal management in the absence of government: a case study in the Dominican Republic. *Ocean and Coastal Management* 36(1-3):47-72.

115. **Viana, V.M., Ervin, J., Donovan, R.Z., Elliott, C. and Gholz, H.** (1996) *Certification of Forest Products: Issues and Perspectives.* Island Press, Washington D.C., USA. 261pp.

116. **McGinn, A.P.** (1998) *Rocking the Boat: Conserving Fisheries and Protecting Jobs.* Worldwatch Paper No. 142. Worldwatch Institute, Washington D.C., USA.

117. **WWF** (1998) *The Final Report of the WTO Shrimp-Turtle Panel: WWF's Response.* WWF Legal Briefing, WWF International, Gland, Switzerland. **Joseph, J.** (1994) The tuna-dolphin controversy in the eastern Pacific Ocean: biological, economic, and political impacts. *Ocean Development and International Law* 25(1):1-30. **Marine Mammal Commission** (1996) The Tuna-Dolphin Issue (pp 99-109 in *Annual Report to Congress for 1995*), Washington D.C., USA. 235pp.

118. **Gjerde, K.M. and Ong, D.M.** (1995) *Marine and Coastal Biodiversity Conservation: the International Legal Framework.* WWF International Discussion Paper, Gland, Switzerland.

119. **Hyvarinen, J.** (1998) *WWF Introductory Guide to the UN Convention on the Law of the Sea (UNCLOS).* WWF International, Gland, Switzerland. 11pp. **Gjerde, K.M. and Ong, D.M.** (1995) *Marine and Coastal Biodiversity Conservation: the International Legal Framework.* WWF International Discussion Paper, Gland, Switzerland.

120. **Fontaubert, C. de, Downes, D. and Agardy, T.** (1996) *Biodiversity in the Seas – Implementing the Convention on Biological Diversity in Marine and Coastal Habitats.* IUCN, WWF, CIEL, Gland, Switzerland. 82pp.

121. **Fontaubert, C. de, Downes, D. and Agardy, T.** (1996) *Biodiversity in the Seas – Implementing the Convention on Biological Diversity in Marine and Coastal Habitats.* IUCN, WWF, CIEL, Gland, Switzerland. 82pp.

122. **Munro, J.L. and Fakahau, S.T.** (1993) Management of Coastal Fishery Resources. Chap. 3. In: Wright, A. and Hill, L. (eds) *Nearshore Marine Resources of the South Pacific.* pp 55-72. Institute of Pacific Studies, Suva/Forum Fisheries Agency, Honiara/International Centre for Ocean Development, Canada.

123. **Kelleher, G., Bleakley, C. and Wells, S.** (1995) *A Global Representative System of Marine Protected Areas. Vols 1-4.* World Bank/IUCN/Great Barrier Reef Marine Park Authority, Queensland, Australia.

124. **Hails, A.J.** (ed) (1996) *Wetlands, Biodiversity and the Ramsar Convention.* Ramsar Convention Bureau/ Ministry of Environment and Forests, India. 196pp. **Thorsell, J., Ferster Levy, R. and Sigaty, T.** (1997) A Global Overview of Wetland and Marine Protected Areas on the World Heritage List. A Contribution to the *Global Theme Study of World Heritage Natural Sites.* Working Paper 2. UNESCO/IUCN.

125. **Spalding, M.D. and Grenfell, A.M.** (1997) New estimates of global and regional coral reef areas. *Coral Reefs* 16:225-230.

126. **Kelleher, G., Bleakley, C. and Wells, S.** (1995) *A Global Representative System of Marine Protected Areas. Vols 1-4.* World Bank/IUCN/Great Barrier Reef Marine Park Authority, Queensland, Australia. **Olson, D.M. and Dinerstein, E.** (1997) *The Global 200: a Representation Approach to Conserving the Earth's Distinctive Ecoregions.* WWF-US, Washington D.C., USA. **Sherman K. and Laughlin, T.** (eds) (1992) *The Large Marine Ecosystem (LME) Concept and its Application to Regional Marine Resource Management.* IUCN, Gland, Switzerland, and Cambridge, UK.

127. **Roberts, C.M.** (1997) Connectivity and management of Caribbean coral reefs. *Science* 287:1454-1457. **Ballantine, W.J.** (1998) *A system of marine reserves: the opportunity and the obligation.* Paper presented at 'Sea Views' conference, Wellington, NZ, Feb. 1998.

128. **Wall, E.** (1996) *UNCLOS and protected areas.* Unpublished manuscript prepared for WWF International.

129. **Kelleher, G., Bleakley, C. and Wells, S.** (1995) *A Global Representative System of Marine Protected Areas. Vols 1-4.* World Bank/IUCN/Great Barrier Reef Marine Park Authority, Queensland, Australia.

130. **Agardy, T.** (1995)*The Science of Conservation in the Coastal Zone: New Insights on how to Design, Implement and Monitor Marine Protected Areas.* A Marine Conservation and Development Report, IUCN, Gland, Switzerland. **Kelleher, G. and Kenchington, R.** (1992) *Guidelines for Establishing Marine Protected Areas.* A Marine Conservation and Development Report, IUCN, Gland, Switzerland. 79pp. **Price, A.R.G. and Humphrey, S.L.** (eds) (1993) *Application of the Biosphere Reserve Concept to Coastal Marine Areas.* A Marine Conservation and Development Report, IUCN, Gland, Switzerland. **Salm, R.V. and Clark, J.R.** (1984) *Marine Protected Areas: A Guide for Planners and Managers.* IUCN, Gland, Switzerland.

131. **Aiello, R.L.** (1996) *Successes and Failures of Global Reef Management.* Reef Tourism 2005. Cairns, Australia. 167pp.

132. **Barzetti, V.** (1993) *Parks and Progress: Protected Areas and Economic Development in Latin America and the Caribbean.* IUCN, Gland, Switzerland/Inter-American Development Bank. **Dixon, J., Fallon Scura, L. and van't Hof, T.** (1993) Meeting ecological and economic goals: marine parks in the Caribbean. *Ambio* 22 (2-3): 117-125. **Kalli de Meyer**, Bonaire Marine Park Manager, pers. comm. 12/6/98.

133. **Reeves, R.R. and Leatherwood, S.** (1994) *Dolphins, Porpoises and Whales. 1994-1998 Action Plan for the Conservation of Cetaceans.* IUCN/SSC Cetacean Specialist Group, IUCN, Gland, Switzerland. 91pp.

134. **Reijndeers, O., Brasseur, S., van der Toorn, J., van der Wolf, P., Boyd, I., Harwood, J., Lavigne, D., Lowry, L. and the IUCN/SSC Sea Specialist Group** (1993) *Seals, Fur Seals, Sea Lions and Walrus: Status Survey and Conservation Action Plan.* IUCN, Gland, Switzerland. 88pp.

135. **IUCN/SSC** (1995) *A Global Strategy for the Conservation of Marine Turtles*. IUCN/SSC Marine Turtles Specialist Group, IUCN, Gland, Switzerland.

136. **Wells, S.M. and Barzdo, J.G.** (1991) International trade in marine species: is CITES a useful control mechanism? *Coastal Management* 19:135-154.

137. **Rose, D.A.** (1996) *An Overview of World Trade in Sharks and other Cartilaginous Fishes*. A 'Species in Danger' report, TRAFFIC International, Cambridge, UK.

138. **Hayes, E.A.** (1997) *A Review of the Southern Bluefin Tuna Fishery: Implications for Ecologically Sustainable Management*. TRAFFIC Oceania and WWF, Sydney, Australia. 34pp.

139. **Vincent, A.C.J.** (1996) *The International Trade in Seahorses*. TRAFFIC International, Cambridge, UK. 163pp.

140. **Phillips, C.** (1996) Conservation in Practice: Agreements, Regulations, Sanctuaries and Action Plans. Chap. 18. In: Simmonds, M.P. and Hutchinson, J.D. (eds) pp 447-465. *The Conservation of Whales and Dolphins*. John Wiley and Sons Ltd, New York, USA.

141. **Phillips, C.** (1996) Conservation in Practice: Agreements, Regulations, Sanctuaries and Action Plans. Chap. 18. In: Simmonds, M.P. and Hutchinson, J.D. (eds) pp 447-465. *The Conservation of Whales and Dolphins*. John Wiley and Sons Ltd, New York, USA.

142. **Wells, S.** (1997) *Giant Clams: Status, Trade and Mariculture, and the role of CITES in Management*. IUCN, Gland, Switzerland and Cambridge, UK. 77pp.

143. **Botsford, L.W., Castilla, J.C. and Peterson, C.H.** (1997) The management of fisheries and marine ecosystems. *Science* 277:509-515. **Mooney, H.A.** (1998) Ecosystem Management for Sustainable Marine Fisheries. *Ecological Applications* 8(1), Supplement. 174pp.

144. **Roberts, C.M. and Polunin, N.V.C.** (1991) Are marine reserves effective in management of reef fisheries? Reviews in *Fish Biology and Fisheries* 1:65-91. **Russ, G.R. and Alcala, A.C.** (1996) Do marine reserves export adult fish biomass? Evidence from Apo Island, central Philippines. *Mar. Ecol. Prog. Ser.* 132:1-9.

145. **Bohnsack, J.** (1996) Maintenance and Recovery of Reef Fish Productivity. Chap 11. In: Polunin, N.V.C. and Roberts, C.M. (eds) *Tropical Reef Fisheries*. Chapman and Hall, London, UK. **Roberts, C.M.** (1997) Ecological advice for the global fisheries crisis. *Trends in Ecology and Evolution* 12(1):35-38.

146. **Mace, P.M.** (1997) Developing and Sustaining World Fisheries Resources: The State of the Science and Management. In: Hancock, D.A., Smith, D.C., Grant, A. and Beumer, J.P. (eds) pp 1-20. *2nd World Fisheries Congress proceedings*. CSIRO Australia, Collingwood, Vic., Australia,

147. **WWF** (1997) *Subsidies and Depletion of World Fisheries: Case Studies*. WWF's Endangered Seas Campaign. 136pp. **WWF Endangered Seas Campaign** (1998) *The Footprint of Distant Water Fleets on World Fisheries*. WWF International, Godalming, UK.

148. **OECD** (1996) *Saving Biological Diversity: Economic Incentives*. OECD, Paris, France.

149. **Miller, D.** (1994) Learning from the Mexican Experience: Area Apportionment as a Potential Strategy for Limiting Access and Promoting Conservation of the Florida Lobster Fishery. In: Gimbel, K. (ed) *Limiting Access to Marine Fisheries: Keeping the Focus on Conservation*. CIEL/WWF, Washington D.C., USA. **Amarasingh, U.S. et al.** (1997) Traditional practices for resource sharing in an artisanal fishery of a Sri Lankan Estuary. *Fisheries Science* 9:311-323.

150. **Alverson, D.L., Freeberg, M., Murawski, S.A. and Pope, J.G.** (1994) *A Global Assessment of Fisheries By-catch and Discards*. FAO Fisheries Technical Paper 339. FAO, Rome, Italy.

151. **Clay, J.W.** (1998) Shrimp aquaculture: where are we headed? *Intercoast* 30:16-17.

152. **Colborn, T., Dumanoski, D. and Peterson Myers, J.** (1996) *Our Stolen Future*. Abacus, UK.

153. **Hinrichsen, D.** (1998) *Coastal Waters of the World: Trends, Threats and Strategies*. Island Press, Washington D.C., USA.

154. **Colborn, T., Dumanoski, D. and Peterson Myers, J.** (1996) *Our Stolen Future*. Abacus, UK.

155. **WWF-Canada and WWF-US** (1998) *Resolving the DDT Dilemma*. WWF-Canada, Toronto and WWF-US, Washington D.C., USA. 52pp.

156. **WWF** (1996) *A Framework Convention for the Phase-out and Elimination of POPs*. A WWF International Position Paper.

157. **Campbell, V.** (1996) *Pollution Prevention: Better Safe than Sorry!* WWF International, Gland, Switzerland. 31pp.

158. **WWF-Canada and WWF-US** (1998) *Resolving the DDT Dilemma*. WWF-Canada, Toronto and WWF-US, Washington D.C., USA. 52pp.

159. **WWF** (1994) *The Right to Know: The Promise of Low-cost Public Inventories of Toxic Chemicals*. WWF-US, Washington D.C., USA. 42pp.

160. **Gjerde, K.M. and Ong, D.M.** (1995) *Marine and Coastal Biodiversity Conservation:The International Legal Framework*. WWF, Gland, Switzerland.

161. **Gjerde, K.M. and Ong, D.M.** (1995) *Marine and Coastal Biodiversity Conservation:The Iinternational Legal Framework*. WWF, Gland, Switzerland.

162. **WWF** (1995) *UNEP Draft Global Programme of Action for the Protection of the Marine Environment from Land-based Activities: WWF Recommendations*. A WWF International Position Paper.

163. **Myers, J.** (1998) *The Cost of Pollution: A Survey of Valuation Methods and their Uses for Policy*. Macroeconomics for Sustainable Development Program Office, WWF, Washington D.C., USA. 59pp.

164. **Fontaubert, C. de, Downes, D. and Agardy, T.** (1996) *Biodiversity in the Seas – Implementing the Convention on Biological Diversity in Marine and Coastal Habitats*. IUCN, WWF, CIEL, Gland, Switzerland. 82pp.

165. **Shah, N.** (1996) *The journey from Arusha to Seychelles: successes and failures in Integrated Coastal Zone Management in East Africa and island states*. Paper presented at 2nd Ministerial Conference on Integrated Coastal Zone Management in Eastern Africa and Island States, October 1996, Seychelles.

166. **GESAMP** (1996) *The Contributions of Science to Integrated Coastal Management*. FAO, Rome, Italy. **Clark, J.R.** (1992) *Integrated Management of Coastal Zones*. FAO Fisheries Technical Paper 327. FAO, Rome, Italy. 167pp. **Clark, J.R.** (1995) *Coastal Zone Management Handbook*. CRC Press Inc., Lewis Publishers, Boca Raton/New York/London. **Pernetta, J. and Elder, D.** (1993) *Cross-sectoral Integrated Coastal Area Planning (CICAP): Guidelines and Principles for Coastal Area Development*. A Marine Conservation and Development Report, IUCN, Gland, Switzerland. **World Bank** (1993) *Noordwijk Guidelines for Integrated Coastal Zone Management*. **OECD** (1993) *Coastal Zone Management: Integrated Policies; Selected Studies*. OECD, Paris, France. 125pp.

167. **Post, J.C. and Lundin, C.G.** (1996) *Guidelines for Integrated Coastal Zone Management*. Environmentally Sustainable Development Studies and Monographs Series No 9. World Bank, Washington D.C., USA.

168. **UNEP** (1995) *Guidelines for Integrated Planning and Management of Coastal and Marine Areas in the Caribbean*. RCU/CEP/UNEP, Kingston, Jamaica.

169. **UNEP** (1995) *Guidelines for Integrated Management of Coastal and Marine Areas with Special Reference to the Mediterranean Basin*. UNEP Regional Seas Reports and Studies No 161. PAP/RAC (MAP-UNEP), Split, Croatia. 80pp.

170. **Mohamad Nasir Abdul Salam** (1998) *An Overview, Handbook 1. Integrated Coastal and Estuarine Area Management*. WWF – Malaysia.

171. **WWF** (1997) *Linking Tourism and Conservation in the Arctic: Ten Principles for Arctic Tourism; Code of Conduct for Tour Operators in the Arctic; Code of Conduct for Arctic Tourists*. WWF Arctic Programme, Oslo, Norway.

172. **IUCN** (1996) *Tanga Coastal Zone Conservation and Development Programme: Logical Framework Analysis*. IUCN – Eastern Africa Regional Office, Nairobi, Kenya, and Irish Aid, Dublin, Ireland. 95pp. **Meltzer, E. et al.** (1997) *Evaluation of the Tanga Coastal Zone Conservation and Development Programme*. Report on IUCN – Eastern Africa Regional Office, Nairobi, Kenya and Irish Aid, Dublin, Ireland. 95pp.

173. **DHKD-WWF** (1996) *Belek Management Plan*.

174. **Robadue, D.** (ed) (1995) *Eight Years in Ecuador: the Road to Integrated Coastal Management*. Coastal Resources Center, University of Rhode Island, Kingston, R.I., USA.

175. **Gibson, J., McField, M. and Wells, S.** (in press) Coral reef management in Belize: an approach through integrated coastal zone management. *J. Ocean and Coastal Management* (1998).

176. **Margoluis, R. and Salafsky, N.** (1998) *Measures of Success*. Biodiversity Support Program, Washington D.C., USA. **Sorensen, J.** (1997) National and international efforts at integrated coastal management: definitions, achievements, lessons. *Coastal Management Journal* 25(1).

177. **WWF** (1996) *Explaining Climatic Change*. WWF, London, UK.